T0294615

SCIENCE AND AGRICULTURAL DEVELOPMENT

This volume examines the complex relationships between science, agriculture, capitalism, and the state and the ways in which these factors have affected the direction of current and projected research. Employing historical analysis and in-depth interviews with scientists, the authors address basic questions of research policy and practice and assess the impact of agricultural research in shaping the modern world.

SCIENCE AND AGRICULTURAL DEVELOPMENT

edited by
Lawrence Busch

ALLANHELD, OSMUN   Publishers

To James S. Brown
Colleague, scholar, and friend

ALLANHELD, OSMUN & CO. PUBLISHERS, INC.

Published in the United States of America in 1981
by Allanheld, Osmun & Co. Publishers, Inc.
(A Division of Littlefield, Adams & Co.)
81 Adams Drive, Totowa, New Jersey 07512

Library of Congress Cataloging in Publication Data
Main entry under title:

Science and agricultural development.

 Based on a graduate seminar at the University of
Kentucky, spring 1978.
  Includes bibliographical references and index.
  1. Agriculture. 2. Agricultural innovations.
3. Agricultural research. 4. Rural development.
I. Busch, Lawrence.
S493.S35   338.1'8   81-327
ISBN 0-86598-022-5   AACR1

82 83 84 / 10 9 8 7 6 5 4 3 2 1

Printed in the United States of America

Contents

Tables and Figures

Preface

This volume originated out of a rather unusual graduate seminar given at the University of Kentucky in the spring of 1978. Unlike many courses, which explore the already well traveled paths of textbook science, this one was to investigate an area overlooked by social scientists: the role of science in agricultural development. Moreover, it was agreed that each student would attempt to produce a publishable manuscript centered around the general theme. The best papers would be revised and put together in an edited volume introducing social scientists to this crucial though neglected avenue for research.

The very fact that social scientists have neglected the topics explored in this volume is in itself a subject worthy of inquiry. Sociologists of science have been largely concerned with physics, believing it "the model of a modern major discipline." Rural sociologists have been more concerned with the diffusion of the products of science than with reflection upon their origins. Those from the various fields concerned with development and modernization have tended to take the products of science for granted, as part of the process of Western progress. Economists have seen science as a factor of production, a kind of "human capital." Political scientists have preferred to dwell on the issues surrounding "technology transfer." Even those who have examined the "Green Revolution" have tended to focus upon the impact rather than the process of science. As a result, the issues presented here, though relevant to a wide range of practitioners, have been largely unstudied.

What is now before you is an attempt to redress that oversight. No effort has been made to produce a unified view of the topic. Each chapter highlights a different facet of the complex interrelationship between science, agriculture, capitalism, and the state in domestic and international development. We are, however, united in our conviction that the world's need for increasing food supplies produced in an environmentally sound and energy efficient manner make the theme of this volume one that we can no longer afford to ignore.

Many persons are involved in the production of an edited collection. At various times Rosemary Cheek, Patricia Bishop, Janice Taylor, Bobbie Sparks, Lynn Webb, and Debbie Pippen assisted in typing the manuscript. Carol Busch read and corrected an early version of the manuscript. Ann Stockham proofread the manuscript, made many valuable comments, and compiled the index. Many of the librarians at the University of Kentucky libraries were essential in helping to unearth some rather obscure publications from the stacks and archives. Moreover, the Kentucky Agricultural Experiment Station provided support for the research reported in several of the chapters. Of course, the responsibility for any errors or omissions lies with the authors.

Lawrence Busch
April 1981

Introduction

THE INTERNAL DYNAMICS OF THE AGRICULTURAL SCIENCES

From the seventeenth century to the present, scientific development has generally been viewed as independent of societal development. Moreover, with the institutionalization of the doctrine of progress during the nineteenth century, it appeared that there was little that science could produce that would not be of benefit to the larger society. This was particularly true in the United States. As Rosenberg has claimed, "moral and scientific progress did not seem contradictory but, to the ordinary American, inevitably parallel and complementary" (1976:3). As a result, "the entire technology of agriculture was machinelike in its advance. Although at times farmers seemed sluggish in their acceptance of improvements . . . there was no organized resistance of workers to its adoption. The new machines, plants, fertilizers, and all the new developments were looked on as undiluted goods" (McConnell, 1953:14). This faith in science infected the Western world and prepared the way for even more rapid technical change.

Nor was this belief held only by conservatives. Even Lenin was convinced of the overwhelming desirability of modern scientific methods in agriculture. In *Capitalism in Agriculture* he asserted that "all European agricultural statistics show convincingly that the larger the area of farms the greater is the proportion of farms using machinery of all types. The superiority of large-scale farming in this very important respect has been fully established" (1938:219).

The non-Western world has become equally enchanted, although this development occurred somewhat later, by the promise of science. The Meiji government of Japan sent a mission to Europe and the United States in the early 1870s. "It identified the source of high productivity as the application of science and technology developed and diffused by agricultural associations, fairs, schools, and experiment stations. These organizations were considered necessary for agricultural improvements in Japan" (Hayami, 1975:49). As is well known, the program of adoption of Western science and technology—in agriculture as well as in industry—helped to move Japan from an unimportant periphery state to one of the major core powers.

This is not to say that the public faith in scientific progress differed substantially from the self-image held by the proponents and practitioners of agricultural science. To the contrary, it appears that researchers have been and remain the staunchest proponents of this view. This appears to be due, in

1

large part, to the particular way in which agricultural scientists (and scientists in general) have come to view their work.

As Mendelsohn has argued, referring to seventeenth-century science, "The consolidators . . . responded to the political realities of their day and consciously banished from consideration, in their new foundations, those areas [politics and ethics] which would transcend the authority of those they did not wish to offend" (1977:17-18). In short, scientists moved from a broad philosophical interest (to better comprehend the work of God?) to a much narrower vocational interest (Weber, 1946). Yet by so doing they could not banish the world of everyday life, for

> every science presupposes the life-world; as purposeful structures they are *contrasted* with the life-world, which was always and continues to be "of its own accord." Yet, on the other hand, everything developing and developed by mankind (individually and in community) is itself a piece of the life-world: thus the contrast is suspended. But this is only confusing because the scientists, like all who live communalized under a vocational end ("life-purpose"), have eyes for nothing but their ends and horizons of work. No matter how much the life-world is the world in which they live, . . . the life-world [itself] is just not their subject matter (Husserl, 1970: 382-3; see also Lukacs, 1971:132).

In short, agricultural scientists have been able to create an abstract, artificial, scientific world and make it *their* world at the same time that they have, unlike those in the basic sciences, consciously maintained a stance directed toward the transformation of the larger social world. As Mulkay (1976) and Daniels (1967) note, the ideal of "pure" science was used to bolster scientists' demands for autonomy. What is extraordinary about the agricultural sciences is that they have been able to demand and receive autonomy while standing clearly on the "applied" side of the dichotomy. Moreover, they have succeeded in this endeavor for over a century. No mean feat, indeed!

This epistemological and ontological dualism, in which the progress of science is inexorably linear and yet is said to lead to "undiluted goods," has manifested itself as "instrumental rationality" (Habermas, 1970). The vocational interests of scientists are separated from issues current within the world of everyday life on the grounds that science cannot define ends but only means. As a result, much agricultural science is reductionist, philosophically naive (in Husserl's sense)—i.e., it takes for granted the social world within which all scientific endeavor takes place—and blindly committed to "progress" as measured by "efficiency."

The reductionist character of the agricultural sciences is well known (Bunting, 1971:442; Janzen, 1975:103). It is manifested in the high level of specialization currently found in agriculture. The purpose of this specialization is to achieve total control over a small aspect of nature. As Chargaff eloquently puts it, referring to much of the biological sciences: "The wonderful, inconceivably intricate tapestry is being taken apart strand by strand; each thread is being pulled out, torn up and analyzed; and at the end even the memory of the design is lost and can no longer be recalled" (1978:56). Yet, as Berry has noted, this attempt at total control tends to lead to disorder (1977:71). The problems caused by the indiscriminate, widespread use of pesticides is a case in point: in defining experiments to test the effectiveness of pesticides, the life-world was defined out; the effects of the pesticides on people and animals were removed from the field of vision.

Similarly, the Cartesian* goal of absolute knowledge is still considered by some to be within reach. One writer, for example, rhetorically asks, "Can our agriculture engineer in century 21 develop theoretical models that can completely and irrefutably describe hydrologic phenomena?" (Lanham, 1976:34). While the author doesn't foresee such models in the near future, it is clear to him that sooner or later they will appear.

The fundamental faith in progress is so deeply ingrained that one experiment station director was able to argue only a decade ago that "the challenge for agriculture research in the 1970's is simple: do better what we are already doing well" (Wood, 1970:102). While in recent years the spokesmen for the agricultural sciences have been more willing to admit to the existence of undesirable consequences of agricultural research, such consequences are still perceived as part of the inevitable construction of absolute knowledge. One spokesman writes as follows: "We are hopefully at an era in our history when social and economic justice and equality, freedom, and stability have become equally as important as efficiency and progress among our societal goals. The problem becomes one of achieving orderly and equitable social and human *adjustment* to the conditions created by technological advance" (Rossmiller, 1969:4; emphasis added). The author takes as given that certain kinds of new technology will be developed; hence, the problem becomes one of ameliorating the more unpleasant effects of that new technology. A recent defense of agricultural science in the prestigious journal *Science* takes a similar view toward any attempt to guide scientific advance (Just, Schmitz, and Zilberman, 1979:1280). In this way science and technology become inhuman forces, reified and conveniently removed from human responsibility.

Perhaps the most telling critique of agricultural science, however, has been its emphasis on efficiency. While values are reduced to secondary status through the conception of absolute knowledge as "the facts," the quest for the illusive goal of efficiency permits practitioners to avoid considering values. Human values are reduced to economic "value." Efficiency—i.e., short-term profit for the farmer-entrepreneur—becomes the yardstick by which research is measured. Indeed, a rather substantial body of literature on the economic returns to agricultural research has developed (e.g., Arndt, Dalrymple, and Ruttan,

*As Descartes (1956 [1637]:13) argued, if the rules of his method are followed then "there cannot be any propositions so abstruse that we cannot prove them, or so recondite that we cannot discover them."

1977; Evenson and Kislev, 1975; Evenson, Waggoner, and Ruttan, 1979). This utilitarian view of agricultural research, particularly prevalent in agricultural engineering and food technology, appears quite reasonable. Yet it belies "a narrow acceptance of the present structure of agriculture as a given condition which restricts options" (Levins, 1973:523). That is to say, it is strongly supportive of the status quo in its effort to be "useful." Thus it simultaneously perpetuates the status quo and denies alternative possibilities. For example, small-scale farm machinery is not developed because it is "inefficient." The large-scale machinery that is developed increases the gap between small and large farmers and convinces the engineer of the rightness of his assumptions. Moreover, it gradually creates a client group that lends political support to the research enterprise on the grounds of its contribution to increased efficiency. As Ruttan explains:

> Under competitive market conditions the early adopters of new technology in the agricultural sector tend to gain while the late adopters are forced by the product market "treadmill" to adopt the new technology in order to avoid even greater losses than if they retained the old technology. One effect of the treadmill phenomenon is . . . to limit the economic motivation for support of agricultural research to a relatively small population of early adopters of new technology. The early adopters also tend to be the most influential and politically articulate farmers (1978:11).

The fact that efficiency, far from being implicit in the natural order, is a socially constructed facade brought about through myriad government policies, the plans of the agribusiness conglomerates, and the prices the Saudis decide to charge for oil, passes by unnoticed. Efficiency is mystified, reified; it is treated as the paramount goal for agricultural research, overshadowing—indeed, concealing—other possibilities. Seen in this light, it is apparent not only that the agricultural sciences have been focused upon those commodities and problems necessary to the maintenance of the world system but that they are structured so as to systematically exclude many alternatives.

Moreover, this model for agricultural research has been actively exported, particularly in the post-World War II period. Foreign students have made up a large proportion of the student body at American colleges of agriculture since at least the 1930s. More recently, through the construction of national and international agricultural research systems, many third world governments have uncritically embraced the apparent cornucopia of American-style agricultural science. Yet doing so has often resulted in the irony of increased production with more inequitable distribution and in greater dependency upon world markets rather than social and economic development.

THE POLICY CRISIS

From the mid-1930s to the mid-1970s agricultural research in the United States was the recipient of secure annual appropriations from the federal treasury. In recent years, however, these appropriations have been called into question. The so-called Congressional farm bloc has disintegrated as the farm population has dwindled to less than 3 percent of the total. Farmers are no longer a special group of people but merely another interest group. Moreover, the very detachment of the urban population from the food-production process has led them to see food products as merely another group of commodities, industrially produced and stacked neatly on store shelves. The dependence of the food system upon biological processes beyond human control has been obscured. Finally, the large agribusiness corporations have themselves become major research institutions. They now spend over $800 million annually on research (National Science Foundation, 1979:55).

As a result of these changes, public support of agricultural research has been eroding rapidly at the same time that agriculture has become a science-intensive enterprise. Moreover, the private research establishment has focused upon those aspects of research which are proprietary. Basic research, research to address environmental and world food issues, research on the social and economic aspects of agriculture, and research on questions of agricultural policy remain restricted to the public domain. Declining public support is likely to mean that these issues will not be examined even in a limited way. Nor can we return to the agriculture that characterized the nineteenth century. That path is forever closed due to rising population, diminishing land resources, and the lack of farming skills on the part of most Americans. Overcoming the present dilemma requires coming to terms with the current research system as well as exposing it as a socially constructed institutional system subject to pressure from vested interests and affected by the fads and foibles of the times.

In this volume we begin that task by inquiring into the complex relationship among the agricultural sciences, capitalism, and the system of nation-states. In so doing we attempt to provide what must be provisional answers to several questions. First, what has been the broader context within which the agricultural sciences have been embedded? Second, what is the process by which the products of the agricultural sciences are constructed? And, third, how and why have the institutions of agricultural research been exported from the West (and particularly the United States) to the Third World?

REFERENCES

Arndt, Thomas M.; Dana Dalrymple; and Vernon Ruttan, eds. 1977. *Resource Allocation and Productivity in National and International Agricultural Research.* Minneapolis: University of Minnesota Press.

Berry, Wendell, 1977. *The Unsettling of America: Culture and Agriculture.* San Francisco: Sierra Club Books.

Bunting, A.H. 1971. Agricultural Sciences. In *Behavioral Change in Agriculture,* ed. J.P. Leagans and C.P. Loomis, pp. 439-80. Ithaca N.Y.: Cornell University Press.

Chargaff, Erwin, 1978. *Heraclitean Fire: Sketches from a Life before Nature.* New York: Rockefeller University Press.

Daniels, G.H. 1967. The Pure Science Ideal and Democratic Culture, *Science* 156:3783.

Descartes, Rene, 1637. Reprint ed. 1956. *Discourse on Method.* Indianapolis: Bobbs-Merrill Co.

Evenson, Robert E., and Yoav Kislev. 1975. *Agricultural Research and Productivity.* New Haven Conn.: Yale University Press.

Evenson, Robert E.; Paul E. Waggoner; and Vernon Ruttan, 1979. Economic Benefits from Research: An Example from Agriculture, *Science* 205 (14 September): 1101-1110.

Habermas, Jurgen. 1970. *Towards a Rational Society.* Boston: Beacon Press.

Hayami, Yujiro. 1975. *Agricultural Growth in Japan.* Minneapolis: University of Minnesota Press.

Husserl, Edmund. 1970. *The Crisis of European Sciences and Transcendental Phenomenology.* Evanston: Ill.: Northwestern University Press.

Janzen, Daniel H. 1975. Tropical Agroecosystems. In *Food: Politics, Economics, Nutrition, and Research,* ed. Philip Abelson, pp. 103-10. Washington D.C.: American Association for the Advancement of Science.

Just, Richard E.; Andrew Schmitz; and David Zilberman. 1979. Technological Change in Agriculture. *Science* 206 (14 December): 1277-80.

Lanham, Frank B. 1976. Heritage and Horizons—How It Began. *Agricultural Engineering* 57:19-34.

Lenin, V.I. 1938. The Theory of the Agrarian Question. In *Selected Works,* vol. 12. New York: International Publishers Co.

Levins, R. 1973. Fundamental and Applied Research in Agriculture. *Science* 181:523-24.

Lukacs, Georg. 1971. *History and Class Consciousness.* Cambridge, Mass.: MIT Press.

McConnell, Grant. 1953. *The Decline of Agrarian Democracy.* Berkeley: University of California Press.

Mendelsohn, Everett. 1977. The Social Construction of Scientific Knowledge. In *The Social Production of Scientific Knowledge,* ed. Everett Mendelsohn, Peter Weingart, and Richard Whitley, pp. 3-26. Boston: D. Reidel Pub. Co.

Mulkay, Michael. 1976. Norms and Ideology in Science. *Social Science Information* 15:4-5, 637-56.

National Science Foundation. 1979. *Research and Development in Industry, 1977.* Technical Notes and Detailed Statistical Tables. Washington, D.C. National Science Foundation, NSF:79-313.

Rosenberg, Charles E. 1976. *No Other Gods: On Science and American Social Thought.* Baltimore, Md.: Johns Hopkins University Press.

Rossmiller, G.E. 1969. Introduction. In *Fruit and Vegetable Harvest Mechanization: Manpower Implications,* ed. B.F. Cargill, and G.E. Rossmiller, pp. 3-5. East Lansing: Michigan Rural Manpower Center, Report no. 17.

Ruttan, Vernon. 1978. Bureaucratic Productivity: The Case of Agricultural Research. Minneapolis: University of Minnesota Department of Agricultural and Applied Economics, Staff Paper 78-16.

Weber, Max. 1946. Science as a Vocation. In *From Max Weber: Essays in Sociology,* pp. 129-56. New York: Oxford University Press.

Wood, G. Burton. 1970. The Research Challenge. National Association of State Universities and Land Grant Colleges, *Proceedings* 84:102-3.

Part I

Planting for Profits

1

American Capitalism and Agricultural Development

S. Buik Mohammadi

The contemporary American economy is characterized by corporate capitalism. Agriculture has not escaped the impact of corporate capitalism; it has been turned into a corporate enterprise itself. American agriculture, modeled after American industry, has been reorganized in accordance with corporate-capitalist principles. It has become agribusiness. The genesis of this phenomenon lies in the peculiar characteristics of the American economy in the first part of the 19th century and the development of corporate capitalism after the Civil War. It is the aim of this chapter to examine the historical development of American agriculture, to consider the impact of the rise of capitalism on agriculture, and to define some of its major characteristics.

NINETEENTH-CENTURY AMERICAN AGRICULTURE

British colonization had put restrictions on American industrialization in order to confine the colonies to the production of raw materials and to keep them as consumers of British products (Faulkner, 1960). Therefore, independent America had little industry on which to rely. As a consequence, political independence from the British did not necessarily provide economic independence as well. It was not until the foundation of a sound domestic economy that America became truly independent from Britain and started competing with that country. The rise of industry and capitalism in America is one of the most important historical phenomena of the nineteenth century. Its paramount impact on history is undeniable.

The striking new features of the American economy at the begining of the 19th century were the rise of cotton production in the South and the appearance of textile factories in the North. Northern agriculture was defined as self-sufficient farming, oriented toward food production (Faulkner, 1960).

In the South, with plenty of land, cotton proved to be the most profitable product. Industrial progress in the North was insufficient, however, to make America an industrialized nation. Cotton remained the major industry, and America was defined as an agricultural country. Railroads and transportation facilities were built to serve agricultural needs—e.g., to haul crops to market.

Habakkuk (1962), comparing British and American industry in the first half of the 19th century, discusses the problem of labor scarcity in America. Not only did America have a labor shortage, but the large geographical area often made the existing labor force inaccessible. Furthermore, land was plentiful and easily obtained, which made the industrial setting unattractive to labor. As a result, wages were high, and the production and use of labor-saving machinery was the main preoccupation of industrialists. In the beginning the United States was better off than Britain—if not in capital, at least in the availability of resources, the better quality of raw materials, cheaper energy, water power, and an abundance of land. Therefore, American industry was oriented toward labor-saving, capital-intensive technology. It seemed more profitable to devise machines that could be constructed quickly and would save labor than to try to save capital. The substitution of capital-intensive for labor-intensive methods depended upon many factors and was not always successful. Nevertheless, the substitution took place. This led to "a greater inducement to organize . . . labor efficiently"(Habakkuk, 1962:45) and to an increase in marginal productivity. Labor scarcity also initiated and encouraged more careful and systematic investigations into techniques designed to keep labor costs low and profits high. By the mid-19th century, although the necessary labor was maintained with difficulties, capital was flourishing in the North and was making its power felt.

The southern United States inherited a cotton culture from the colonial period. Despite the innovations in labor-saving machinery, the need for these implements was not urgent. Cheap slaves provided the labor necessary for the plantation. What mattered was an increase in the agricultural product, namely cotton. A comparison of United States agriculture to that of England reveals differences in their orientations toward labor and land. "In America improvements in agriculture took the form primarily of increasing output per head . . .[while in England] agricultural improvement was devoted primarily to increasing yield per acre"(Habakkuk, 1962:14). In America there was more than enough land for exploitation. Therefore, "southern wealth was invested primarily in cotton, land, and slaves" (Faulkner, 1960:315). The Southern attitude toward land was not different from the attitude toward slaves. Land was bought, used, and then sold after being exhausted. Land was viewed as a replaceable piece of machinery or another tool (Wasserman, 1972). It is not difficult to see why the South favored the rapid sale of Western land at low prices and was opposed to land restrictions. In contrast, through high land prices, the North favored concentration of the population in the cities and was opposed to the encouragement of westward migration (Faulkner, 1960).

By the eve of the Civil War, the general shape of American agriculture had crystallized: it consisted of land-extensive, mechanized farming methods and the well-defined commercialization of agriculture. The effects of a plantation economy were apparent. The plantation economy was based on continued monocultural production, and capital was usually provided from outside the South—from the North or from England. Such an economy, financially dependent and based on "King Cotton," was susceptible to price fluctuations in the world market. The lack of independent capital and heavy industry pushed the South into an inferior economic position (Faulkner, 1960). Under these circumstances, the results of war and a Northern victory might have been foreseen by an imaginative mind.

The immediate impact of war in the South was a food shortage (Faulkner, 1960; Williams, 1969). Railroads were used for pressing military purposes, which made the food supply problematic. Arguments for extending multicrop production started. More than ever before, the need for agricultural diversification was felt. However, devastation by the war, bad weather, and crop failures left little option. As Williams put it, "the South after its surrender was 'a land of poverty' and 'one of America's colonies' " (1969:108).

The war demonstrated the effectiveness of industrial capitalism and brought the weaknesses of the plantation economy to the surface; the Northern victory revolutionized both industry and agriculture. The abolition of slavery provided additional labor for the factories. Slavery was a violation of the laissez-faire principle and an obstacle to capitalist development and industrialization. The freedom of the slaves provided freedom for industrial capitalism.

The changes in American agriculture during and after the Civil War are as significant as the changes in the realm of industry. In 1862, four important acts were passed: (1) the act that authorized establishment of the United States Department of Agriculture (USDA), (2) the Morrill Act, providing for the development of land-grant colleges, (3) the Homestead Act, and (4) the Pacific Railway Act. These laws were passed without the consent of the South, but they were nevertheless later applied to the South as well as the North. "These... historic acts," in George McGovern's words, "combined with the growing volume of federal farm program legislation and executive action, have played an increasingly determinative role in American agricultural life" (1967:xviii).

A full understanding of American agriculture after the Civil War is possible only by a thorough analysis of the rise of corporate capitalism and the relationship of agriculture to the industrial sector. The Jeffersonian ideal of an agricultural America embodied a romantic symbolism, while agriculture, under the influence of businessmen, increasingly turned into a capitalistic enterprise. Spurred on by the war, the metropolitan minority gradually dominated the agricultural majority. Adam Smith's competitive market and laissez-faire system became the ruling ideology. When capitalism was turned loose, industry replaced agriculture and mercantilism as the basis of the economy. The new mode of production was accompanied by a new mode of social relationships and a new social structure.

Having defined the economic institutions in terms of laissez-faire, the industrial elite entered into the entrepreneurial competitive market. However, given the special advantages available to a few, free competition did not last long. The gospel of free competition exhausted its usefulness, and the need for some kind of consolidation became apparent as early as the 1870s. At least three factors can be counted as responsible for the systematic elimination of competition and the rise of oligopoly in American economic institutions:

1. The capitalist elite realized that competition was an obstacle to progress when it proved to be unprofitable. Capitalists found that by competing with one another they were wasting capital rather than accumulating it. Industrialist J.P. Morgan, the baron of the railroads, talked about the horrors of competition and mistrust and called for a business consolidation euphemistically named the "community of interests" (Williams, 1961). The well-organized, oligopolistic railroad system provided such a model. The impact of the Pacific Railway Act of 1862 on American agriculture was immense. The railroad not only exploited the farmers; it also paved the way for corporate enterprise through the suggestion that other industrialists should organize themselves in the same fashion (Josephson, 1962). In 1899, the Industrial Commission asserted that "competition [was] so vigorous that the profits of nearly all competing establishments were destroyed" (Commission's Report, 1900:9). This was the chief motive for industrial combination.

2. From its inception, American industry tended toward largeness. Partly due to the belief that bigger was always better and partly due to labor scarcity, large mass-producing industry became characteristic of American business. "Thus the history of [industrial] consolidation and of the growth of big business are closely interwoven. The invention of labor-saving machinery made large-scale production profitable, the heavy fixed investment in expensive machinery and apparatus discouraged competition, and the very growth in the size of the nation and its business tended, as in the case of the railroad, to inevitable consolidation " (Faulkner, 1960:422).

3. During the second half of the 19th century, there was much well-organized labor and farmer unrest in America. Competition made it difficult to deal with these disturbances. In order to save the system from radical alteration by industrial labor and farmers, businessmen united. Therefore, the consolidation of business was in part due to the formation of class consciousness among businessmen. As McConnell put it, "By conquest and use of governmental power, the capitalists had fastened a new tyranny upon the common people, that is, upon farmers. The interest of a rising and powerful faction of a few rich men thus stood in direct opposition to the common interest; it was a class struggle between capitalists and agrarians" (1953:7).

In changing from laissez-faire to corporate capitalism, business went through various intermediate stages (Faulkner, 1960). By the turn of the century, the final stage had been reached and the general characteristics of the corporation in capitalist enterprise were established.

The half century between 1860 and 1910 is considered by many historians to be the first agricultural revolution in the United States. In some sense this is true; the events that took place during this period laid the foundation for contemporary American agriculture. Expansion west of the Mississippi, free land, and the Homestead Act were issues long before the Civil War. In fact, in 1860 the Homestead Act was vetoed by President James Buchanan and called "communistic" (Wasserman, 1972). Even though both the South and the North wanted expansion beyond the Mississippi River, the Homestead Act did not become a law until 1862. The problem was not whether to expand, but expansion on whose terms. The South wanted the western land for plantations and slavery (Wasserman, 1972); the North wanted the West to be turned into a capitalistic enterprise that would make use of its resources.

The Homestead Act provided to American citizens "160 acres of public domain to any person who was the head of a family or over 21" (Rasmussen, 1960:112). However, from the beginning the act proved to be inadequate. The very content of the act and its application reveal that the pattern was set under the metropolitan industrialist's terms. On the surface, the act seemed to be a good idea; in practice, the free land and farming proved to be neither free from charge nor free from problems: "Only one of every ten families that went West after the Civil War actually got free land. High taxes—most of which went to pay off money grants to the railroad barons—and high mortgage and interest rates made the load even heavier. It no longer took so much courage, strength, ability to start a farm—it took $1,000" (Wasserman, 1972:64). Gates (1936), discussing the misinterpretation regarding the Homestead Act, makes the case that the "free land" turned into a profitable business. First of all, the railroad managed to get the most desirable land possible. Indeed, it created value by placing a railroad on a given tract. Second, the speculators secured what was left of the desirable land. Finally, the timber dealers soon possessed the richest woodlands. As a result, farmers were turned into tenants. The free land was that which neither the land speculators nor the railroads were interested in— the least desirable land. Besides this, the impact of "cheap and free land" has been greatly exaggerated. The cost of farming itself was increasing. It has been estimated that between $750 and $1,500 was required to bring a prairie farm into production (Gates, 1965). Furthermore, titles to the land were often stolen from the farmers; businessmen poured into Western homestead territories, engaging in fraud, extortion, and land speculation. "Soon the farmers found good farming and dollar solvency were two different things" (Wasserman, 1972:64).

The slaves, having been freed and turned into "free labor," added to the concentration of the population in the cities and metropolitan areas. At the same time the number of farms increased. Thus, the problem of farm labor scarcity after the Civil War was amplified. The solutions for the problems of labor scarcity and the lack of settlers were provided by the Northern capitalists: the Contract Labor Law of 1864 was enacted in order to increase the flow of immigrant labor (Williams, 1961). Advertisements in Europe invited people to America, offering free transportation and making glorious promises. Labor scarcity was also reduced through the creation of labor-saving machinery. This was brought about through more extensive use of already existing labor-saving implements like the plow and the reaper, and an extensive replacement of men by horses as a source of power in agriculture. For example, the superintendent of the census in 1869 argued that, because of high labor costs, American agriculture was not profitable and European experiments could not be applied in the United States. He urged the extensive use of labor-saving implements and the replacement of wagons by the railroad (Rasmussen, 1975). The North had aleady set the example. When the war drew half a million men from the farms, their places were filled by farm implements and machinery (Rasmussen, 1975). Between 1861 and 1865 the production of mowing machines, for example, increased 250 percent, and farmers purchased 250,000 reapers and mowers (Williams, 1969). Furthermore, innovative large machinery was introduced, such as the twine-binder and the combine, which was propelled at first by horses and later by gasoline (Faulkner, 1960).

The need for labor-saving machinery in agriculture encouraged the development of large-scale capital-intensive implements. So enormous were these machines that they were frequently referred to as "leviathans" (Wik, 1975). Their high cost put them out of the reach of small farmers. Furthermore, having high capacity, the machines were not suitable for small farming operations. The gigantic size of farm machinery permitted what has frequently been referred to in the literature as "bonanza farming." William G. Moody (1968), describing its impact in 1883, noted that farm size had increased and that even larger farm operations had become possible under the power of machinery and capital. From 100 acres before agricultural machinery and the railroad, farms grew to 1,000 acres, then to 10,000 acres, and at the time of his book they were still increasing in size. He observed that, for example, the Rock County Farming Company contained 21,000 acres, of which 4,625 were under cultivation. There were 96 horses and mules, 26 harvesters, three straw-burning thrashers, and other farm implements on the farm, valued up to $15,000. The implications were obvious: "This state of things is made possible, and is obtaining, [*sic*] solely by and under the power and use of machinery; first in the hands of individual capitalists; then in the hands of companies; and lastly, by corporations" (Moody, 1968:59).

In California other innovations in machinery took place. For example, the Moline plow and sulky cultivator were combined into a gang-plow, and a gigantic header replaced the self-binding reaper. An 1878 USDA annual report suggested that "a new science will have to be set up to meet the case of California" (Rasmussen, 1975: 1479).

One of the main problems of the farmers was transportation. The metropolitan industrial capitalists provided the "solution" for this problem, too. The Pacific Railway Act of 1862 gave the promoters of the railroads (specifically the Union Pacific and Central Pacific) five square miles of land for every mile of track they would lay across the West. Two years later, the Congress decided to increase the amount of land to ten square miles in addition to an allotment of up to $40,000 for every mile of track. In ten years, the railroad companies owned $700 million and 200 million acres of public land (Wasserman, 1972).

Railroad involvement in the transportation of agricultural products is one of the classic examples of exploitation and control in American history. The statistical data show the steady increase in capital equipment investment in the railroad industry, particularly after the Civil War, when the railroad moved westward (see Table 1.1).

Table 1.1   Total Stock, Mortgage, Bonds, Equipment, and Obligations and Total Traffic Economy of Railroad Companies, Selected Years

	Total stock, mortgage, bonds, equipment, and obligations	Total traffic economy (in thousands of dollars)
1867	1,172,881	334,000
1871	2,664,628	403,329
1876	4,468,592	497,258
1881	6,278,565	701,781
1886	8,163,149	829,941
1890	10,122,636	1,097,847

Source: U.S. Bureau of the Census, Historical Statistics of the United States, Colonial Times to 1957 (Washington, D.C.: U.S. Government Printing Office, 1961), p. 428.

There were, of course, times when the railroad rates were reduced, but the costs were never low. Transportation costs were one of the main expenses of farmers. "In some areas the freight charges incurred in moving crops to a market might absorb as much as the crop's value at that market" (Higgs, 1970:291). In other words, farmers often found themselves having to pay the price of a bushel of wheat in order to have another bushel shipped to the market (Wasserman, 1972). Moreover, the transportation cost for small farmers was often 50 percent higher than the rate charged for large farmers

(Moody, 1968). This situation explains why railroad exploitation was one of the main causes of the Populist movement in the late 19th century. The Populists had rightly recognized that a great part of their hardship was caused by freight rates and the railroads (Higgs, 1970).

Another problem of the farmers was the steady decline and fluctuation of farm prices. The price of industrial products either did not decline or did not decline as much as farm products. The result was pressure on the farmers to quit, to move elsewhere, or to be a tenant farmer or rural laborer (Faulkner, 1960).

The irony was that production increased. Farmers produced more and earned less. Production as a whole, production per acre, and production per laborer all increased; yet farmers became destitute. Of course, in times of depression and crisis, the prices of other goods, materials, and services declined, too, but not to the same extent as farm products. For example, in the "Great Depression" of the mid-1890s, the transportation rate declined 10 to 15 percent, but farm prices dropped 30 to 50 percent (Higgs, 1970). The expense of purchased inputs was also frequently a burden on farmers. For example, during the 1890s, fences were among the most expensive of farmers' needs. Regardless of the type of fence, famers realized that "for every dollar invested in livestock, another dollar is required for the construction of defenses to resist their attack on farm production" (Rasmussen, 1975:1402).

In short, the prices of the industrial products upon which farmers were increasingly dependent either stayed high or declined little compared to the decline of farm product prices. The lack of control over prices, not only of the products they consumed (implements, etc.) but also of what they produced, pushed farmers into a powerless position. "Technology freed farmers somewhat from the blind forces of nature and made them victims of the equally blind forces of market fluctuation. The term 'farm problem' became synonymous with 'farm surplus' " (McConnell, 1953:14). Farmers felt the full impact of capitalism. Cattle and corn farmers could get probably half the value of their product; and wheat farmers, if lucky, got about one-fifth of the value of their production. "In the depression of the nineties, while crowded cities starved, Westerners burned corn for fuel because it was worth less than coal" (Wasserman, 1972:66).

The farmers were caught among many evils: expensive large agricultural machinery, the expense of irrigation and irrigation implements, lack of sufficient capital to meet these expenses, high transportation costs, discrimination, misuse of the railroads, and overproduction and lack of control over the prices of farm products. All of these problems, directly or indirectly, discouraged small-scale farming and favored large-scale land ownership (Gregor, 1970). Farmers had to deal with the risks of cultivation and take the responsibility for loss. Their lives were invested in the land, yet the land could be taken away from them if they had a bad season. For example, "after the drought of 1887, at least 11,000 Kansas families were evicted. Twenty towns in the Western part of the state were left to the ghosts" (Wasserman, 1972:65).

In theory, farming meant freedom, self-sufficiency, and progress. In reality, farming meant gambling. In other words, American *agriculture* proved to be a flourishing business, but farming did not. The urban-based economy commodified whatever aspect of agriculture happened to be profitable and extracted the surplus from the farmers, as had been done with the urban proletariat. In Williams's words, the American farmers

> observed the ground rule of *laissez-faire,* applied the new technology, used the new machinery, specialized in regional crops, and produced more—yet their share of the system's income decreased. Freight rates and other industrial prices fell about 67 per cent between 1865-1896, but farm prices dropped about 75 per cent. During the same years, moreover, the exportable surplus of wheat jumped 16 per cent. And by 1885, even the federal land commissioner acknowledged that the public domain was being made prey of unscrupulous speculation and the worst forms of land monopoly through systematic fraud (1969:335)

The angry farmers' opposition to deflation, the discriminatory credit system, monopolies, railroad abuse, land speculators, and middlemen did not change the trend. By the close of the century, the pattern was fixed. The nonfarm aspects of agriculture were dominated by the corporations and businessmen. And the hardships of farming were left to the farmers. Agriculture was turning into agribusiness. The snowball had already formed and started rolling.

CORPORATE FARMING AND AGRIBUSINESS

Agriculture has not escaped the effects of corporate capitalism. Agricultural problems and issues are defined within a capitalist framework, and solutions are offered by capitalists in favor of mighty businessmen. The result is the inevitable formation of corporate farming and agribusiness.

"The term 'agribusiness' is used to denote large, industrial-type operations along the commercial food chain,[and the term] 'corporate farming' sums up images of big business owning and cultivating limitless acreages" (Walsh, 1975:29). In other words, "corporate farming" refers to raising crops and livestock on a large scale, while "agribusiness" means capitalist, industrial, and financial involvement in nonfarm agriculture. Agriculture in this sense includes a large variety of activities from farm-implement manufacturing to marketing, storage, and processing of food. The lines, of course, are not clearly drawn. It is possible that a corporation involved in farming may be involved in other agricultural activities as well. This is probably the reason some authors use the terms "corporate farming" and "agribusiness" inter-changeably.

Both corporate farming and agribusiness have put farmers in an inferior and subordinate position. According to Goss and Buttel, " 'dominance' in agriculture, even in 1900, was no longer in the hands of farmers, but rather was moving toward the railroad, machinery manufacturing, and food processors.... Farmers have hardly been insulated from the forces which generate large-scale agriculture" (1977:18). Let us examine each of these trends separately.

Corporate Farming

The literature on corporate farming is broad, particularly the popular literature concerning the small family farm versus the corporate farm. Apart from the controversies over definitions of corporate farming or the interpretation of the data, the fact is that some important changes in American agriculture have taken place, "of which the possible increase in corporate farms is only one" (Goss and Rodefeld, 1977). The general characteristics of corporate farms have been summarized as follows:

—increased scale of farm operations
—increased total farm production
—increased specialization in type of farm production
—increased vertical integration through production and price contracts
—increased use of leasing, custom work and credit arrangements
—increased nonfarm ownership of land and other farm resources
—decreased number of farms
—decreased number of farm owner-operators
—increased proportion of full-time hired farm laborers and managers
—increased part-time farming
—decreased political power of farm people (Goss and Rodefeld, 1977:44)

Probably among the most frequently mentioned characteristics of American agriculture are the vanishing small farm and the growth of big farms (Barnes, 1971; Walsh, 1975). The family-versus-corporate-farm debate has been prominent from time to time; it was particularly manifest during the Great Depression (Goss and Rodefeld, 1977). The debate continues and the case has been made that the vanishing of small farms is now more of a problem than ever before.

Barnes (1971) points out how difficult it is for small farmers to survive and how easy it is for corporations to succeed. Corporations have the advantages of (1) the ownership and control of large machinery, (2) the ownership of agricultural chemicals and the packaging and marketing process, (3) access to easy credit, (4) large benefits from government in the form of subsidies, (5) subsidies in the form of water and nonenforcement of federal water laws, and (6) the cheaper input costs of bulk buying. By the same token, the small farmer has the disadvantages of (1) competition with corporations, (2) lack of access to easy credit, (3) often, lack of access to and ability to hire farm labor, and (4) in some cases, lack of water because of the presence of corporations. That is why corporate farming proves to be profitable.

Moreover, corporate farming is more than just food production. It is "land speculation, tax dodging, and the development of an integrated total food system" (Barnes, 1971:22). There is little wonder that only a few people without other sources of income can enter into farming today. The result is apparent: small farmers quit, and big farms get bigger. According to USDA statistics, in 1954 there were 5.4 million farms in America. By 1977, this number had declined to 2.9 million. Meanwhile, the average size of a farm over the last 20 years increased from 215 to 380 acres (Barnes, 1971).

Agribusiness

Agribusiness involves the entry of large oligopolistic corporations into the agricultural sector. A monopoly may involve vertical or horizontal integration. Horizontal integration implies the monopoly of the manufacturing and sale of a commodity by a few corporations. Vertical integration means that corporations have sole control "over the companies that feed it—raw material, component parts . . . [and] wholesale or retail" (Tyler, 1976:79) trade. The advantage of vertical integration is that the corporation can avoid any business process that is unprofitable or risky, control whatever is profitable, and increase control over production. In agriculture, corporations can leave the family farmers alone, letting them do the farming but controlling the other aspects—e.g., the input and output segments—of agriculture. "In doing so, it [the corporation] gains many of the advantages of family farming (primarily cheap labor) without taking from the farmer the risk of ownership" (Tyler, 1976:101).

The history of vertical integration goes back to sugar-beet cultivation in the 19th century. In the 1870s, California became a successful pioneer in the sugar-beet industry. "The factory inaugurated the practice of contracting with farmers for a specific acreage of sugar beets" (Rasmussen, 1960:129). Later this practice became a model for other agricultural industries.

The data show that farming is an inferior business in comparison to the output and input segments of agriculture. As Figure 1.1 indicates, the output-input sectors of agribusiness are larger operations and bigger industries, both in labor usage and in a variety of other activities.*

It is difficult to separate farm problems from the issues of corporate farming and agribusiness. As the "mortality" of the small farmer increases, the size of big farms increases and the number of farms decreases. Concentration, centralization, consolidation, and expansion are parallel phenomena. An outcome of the specialization in the production of inputs and outputs is specialization of the farm product (Walsh, 1975).

This process increases the power of corporations and forces farmers into

*It is difficult to update the data in Figure 1.1. No direct data could be found on agribusiness employees. Snodgrass and Wallace have probably used their own estimations. Nevertheless, there is good reason to speculate that the agribusiness complex is growing while farm activity is declining. In 1977, 4,375,900 people were involved in farming (USDA, 1978), compared to 6 million people in 1970.

Figure 1.1   Scope of Modern Agriculture

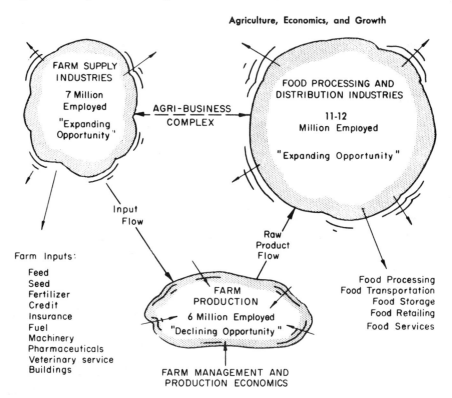

Source: Milton M. Snodgrass and Luther T. Wallace, *Agriculture, Economics and Growth* (New York: Appleton-Century-Crofts, 1970), p. 12

even fiercer competition. Even though food prices rise, the prices of farm products remain low. There is one simple explanation: "The profit is not, in general, to be found in the production end of the business" (Walsh, 1975:29). According to Zwerdling (1976), agribusiness—or, using his terminology, the food monopolies—are the fundamental source of the food crisis in America. He demonstrates how the USDA's report of rising food prices is accompanied by the growing profits of chain food stores. Safeway boosted its profits 51 percent in nine months; Kroger increased its profits 94 percent. What Barnes (1971) calls the "integrated total food system" Zwerdling labels a "corporate empire." Here is an example of one such mighty empire:

> At Safeway's latest count, this $7 billion food empire owned more than 2,400 supermarkets, 109 manufacturing and processing plants, 16 packaging plants, 16 bakeries, 19 milk . . . plants. The Safeway corporate strategies are formulated by the

board of directors, titans of industry who also help control such corporations as the Bank of California, General Electric, . . . Shell Oil, etc. (Zwerdling, 1976:44).

While the USDA is preoccupied with price fluctuations, analyzing the supply and demand curves, giant corporations have enough control to be able substantially to influence supplies and prices. At the production end of the food chain, control is not as extensive as in some other American industries; but in the retail market—"where the concentration really counts—much of the food industry is revealed as even more tightly controlled than energy" (Zwerdling, 1976:46). Total control of a given agricultural product is probably impossible; yet any corporation controlling 50 percent of the market for a given product can control prices easily. Big corporations can crush small companies in a short time by the selective manipulation of prices (Zwerdling, 1976).

The position of farmers in this situation is a frustrating one. The high productivity of American agriculture is undeniable. Even though the acreage in farm land is stable, economic indicators reveal the rising productivity of American agriculture. Sometimes the literature on American agriculture even talks about the United States feeding the world. The inevitable question comes up, where does the American farmer stand in relation to such a productive agriculture?

In the 20th century, U.S. agriculture has seen three full-scale revolutions: "mechanical, technical, and business management" (Tyler, 1976:99). Each revolution introduced some innovations; as a result, the importance of new implements, new chemicals, and new management rose to a point where they became imperative to survival in American agriculture. In other words, increasingly corporate capitalism has revolutionized agriculture and made the division between the farm and nonfarm activities of agriculture sharper than ever before. For example, "in 1940, two-thirds of all input was traditional—labor and land; only one-third was nonfarm, like buildings, machinery, fuel oil, fertilizer, . . . outside labor" (Tyler, 1976:100). Twenty years later this proportion was reversed. More and more farmers have realized that large-scale farming and large-scale usage of agricultural technology are most profitable. In small-scale agriculture, the human and technological potential remain underutilized. Therefore, "in agriculture as in industry, technical progress and increase in labor productivity have made production steadily more capital-intensive" (Averitt, 1968: 165).

The results of these revolutions in agriculture were twofold. First of all, those who could not keep up with the "modern" methods of production were wiped out (Tyler, 1976). Second, farmers became dependent upon corporations for capital inputs and capital-intensive output processes. In Averitt's words, "the financial rewards in small-scale farming have flown away, lifted on the wings of technological change toward the supplier of capital and marketing technique" (1968:166).

A good example of this process is tomato harvesting in California, which is characterized as an assembly-line operation (Friedland and Barton, 1975). The results of such an operation have been (1) high production, (2) concentration of tomato processing in California, (3) a geographical shift in tomato production due to new harvesting machines, (4) development of a successful price-bargaining position in California, and (5) a change in the characteristics of harvest labor forces from migratory Mexicans to settled Americans.

Since oligopolistic control has increased in both the input and output markets, the farmers find themselves dependent upon corporations on both sides. Therefore, the magnitude and type of production depends upon the willingness and availability of corporate facilities to pack and to ship the product to the market, and the corporation can decide both the price of farm products and the cost of processing and marketing.

The oligopolistic character of American agriculture, despite its advantages and technical progress, has left American farmers in an economically stagnant position (Averitt, 1968; Tyler, 1976). The centralized input-output dimensions of agriculture and the growing importance of the nonfarm sector overshadow actual farming activity. As a result, farmers are forced to compete in a contrived "free" market. Not only does the high cost of inputs reduce the possible profitability of farming, but also farmers are subject to low prices for their products. "Farmers are archetypal price takers" (Averitt, 1968:161). Any price fluctuation touches the farmer more than anyone else involved in agriculture. Farmers, through painful experience, have learned that high production does not necessarily mean economic prosperity. In other words, farmers are engaged in the least profitable and the most competitive segment of agricultural enterprise. The farmer is but cheap labor exploited by the corporations involved in agribusiness.

The capitalist mode of production in agriculture goes beyond food raising. Indeed, the invasion of farm activities by nonfarm corporations overshadows the food-raising activity. Activities like processing, packing, and marketing become the main concern of agribusiness, persisting even if the land monopoly disappears. The price-making by the non-food-raising sectors of agriculture causes a falling rate of profit on the farm (Goss and Buttel, 1977). The nonfood-raising processes become so important and so divorced from farm and food-raising activity that they make agriculture dependent on the nonfarm sector. More than in any other period in history, the American countryside has become dependent upon the town. As Goss and Buttel put it, "The increasing specialization of farmers in food raising (and the corresponding monopoly capitalist penetration of inputs, processing, and marketing) also creates the condition under which the farmer becomes subject to exploitation and monopoly control by nonfarm segments of agriculture" (1977:21).

Specialization in food production by farmers and by region decreases the number of farms and farmers. More and more, application of huge machinery makes farming a specialized profession and makes farmers dependent upon corporations. The growing importance of nonfarm activities has put farming in an inferior position. The result is a corporate-capitalist mode of production that contains in itself the basic problems and contradictions of capitalism.

From the point of view of agribusiness corporations, this mode of agricultural production is *Paradise Regained;* it is stable and highly controlled. From the point of view of farmers and consumers, with constantly rising food prices in spite of the low prices of farm products, it is highly unstable. Specialization and assembly-line mass production leave agriculture open to the influences of unpredictable factors. Bad weather in California is enough to destroy the tomato crop, as it has that of lettuce. Even though they are food producers, farmers, like other consumers, are dependent upon corporations for their food. The corporate structure of the American economic system controls agriculture, and this control has turned farmers and nonfarmers alike into powerless masses who pay the price of empire building, while the ideology of free enterprise, which lost its usefulness a hundred years ago, still haunts America.

REFERENCES

Averitt, Robert T. 1968. *The Dual Economy; the Dynamics of American Industry Structure.* New York: W.W. Norton & Co.
Barnes, Peter. 1971. The Vanishing Small Farm. *New Republic* (12 June) 164:21-24.
Commission's Report. 1900. Preliminary Report of the Industrial Commission on Trusts and Industrial Combinations, vol. I.
Faulkner, H.U. 1960. *American Economic History.* New York: Harper and Brothers.
Friedland, William H., and Amy Barton. 1975. *Destalking the Wily Tomato: A Case Study in Social Consequences in California Agricultural Research.* Davis: University of California, Dept. of Applied Behavioral Sciences, Monograph #15.
Gates, Paul Wallace. 1936. The Homestead Law in an Incongruous Land System. *American Historical Review* 41, no. 4 (July):652-81.
―――――. 1965. *Agriculture and the Civil War.* New York: Alfred A. Knopf.
Goss, Kevin F., and Frederick H. Buttel. 1977. The Political Economy of Class Structure in Capitalist Agriculture. Unpublished paper.
Goss, Kevin F., and Richard D. Rodefeld. 1977. *Corporate Farming in the United States: A Guide to Current Literature, 1967-1977.* Department of Agricultural Economics and Rural Sociology. The Pennsylvania State University.
Gregor, Howard F. 1970. The Industrial Farm as a Western Institution. *Journal of the West* 9, no. 1:78-92.
Habakkuk, H.J. 1962. *American and British Technology in the Nineteenth Century: The Search for Labor-Saving Inventions.* Cambridge: Cambridge University Press.
Higgs, Robert. 1970. Railroad Rates and the Populist Uprising. *Agricultural History* 44, no. 3 (July):291-97.
Josephson, Mathew. 1962. *The Robber Barons: The Great American Capitalists 1861-1901.* New York: Harcourt, Brace and World.
McConnell, Grant. 1953. *The Decline of Agrarian Democracy.* Berkeley: University of California Press.

McGovern, G. 1967. *Agricultural Thought in the Twentieth Century.* New York: Bobbs-Merrill Co.

Moody, W.G. 1968. Land and Labor in the United States. In *The Transformation of American Society, 1870-1890,* ed. J.A. Garraty, pp. 41-48. Columbia: University of South Carolina Press.

Rasmussen, Wayne D. 1960. *Readings in the History of American Agriculture.* Urbana: University of Illinois Press.

———, ed. 1975. *Agriculture in the United States; A Documentary History.* Vol. 2. New York: Random House.

Snodgrass, Milton M., and Luther T. Wallace. 1970. *Agriculture, Economics and Growth.* New York: Appleton-Century-Crofts.

Tyler, Gus. 1976. *Scarcity: A Critique of the American Economy.* New York: New York Times Book Co.

U.S. Bureau of the Census. 1961. *Historical Statistics of the United States, Colonial Times to 1957.* Washington, D.C.: USGPO.

USDA. 1978. *Agricultural Statistics, 1978.* Washington, D.C.: USGPO.

Walsh, John. 1975. U.S. Agribusiness and Agricultural Trends. In *Food, Politics, Economics, Nutrition and Research,* ed. Philip H. Abelson, pp. 29-32. Washington, D.C.: American Association for the Advancement of Science.

Wasserman, Harvey. 1972. *Harvey Wasserman's History of the United States.* New York: Harper & Row.

Wik, Reynold M. 1975. Some Interpretations of the Mechanization of Agriculture in the Far West. *Agricultural History* 49, no. 1 (January):73-83.

Williams, William Appleman. 1961. *The Contours of American History.* Cleveland: World Publishing Co.

——— . 1969. *The Roots of the American Empire: A Study of the Growth and Shaping of Social Consciousness in a Marketplace Society.* New York: Random House.

Zwerdling, Daniel. 1976. The Food Monopolies. In *Crisis in American Institutions,* ed. Jerome Skolnick and Elliott Currie, pp. 43-51. Boston: Little, Brown & Co.

2

From Farmers to Miners: The Decline of Agriculture in Eastern Kentucky

Sally Maggard

During a period of about five decades, roughly between 1880 and 1930, a self-sufficient farm population in eastern Kentucky was transformed into an industrial work force. In the process, a farm economy was destroyed. This chapter focuses on how the agricultural sciences as they were developing in the United States, and particularly in Kentucky, figured in the demise of mountain farming.

Eastern Kentucky has long been a leader in the United States in coal production. It was almost totally agricultural prior to coal and timber exploitation. During the years of the area's transformation, the groundwork for the University of Kentucky's College of Agriculture was being established. Eastern Kentucky, therefore, offers an important case study in any historical analysis of the orientation of university agricultural research and extension work.

EASTERN KENTUCKY: THE INDUSTRIALIZATION OF A RURAL REGION

Eastern Kentuckians are generally thought of as "hillbillies"—as rather crude, unpolished people living in the Appalachian mountains. The two most common images of mountain people are of coal miners and of farmers. The first image has the feeling of present tense. It is associated with energy needs, electricity, union troubles, and periodic mine-related disasters. A decade ago, in the 1960s, there was also an awareness of coal mines shutting down, high unemployment, and stark poverty in mountain hollows.

The second image—of mountain farmers—has the feeling of past tense, of nostalgia. People think of pioneer settlers and scenic trails. In fake log cabins along scenic routes, tourists can buy relics from a long-dead agricultural way of life. Sadly, descendants of these dispossessed farmers are often portrayed as comical and lazy moonshiners popping up out of cornfields to crack jokes. Even worse, there is the image of a primitive near-beast, aiming his rifle at any foreigner who happens across his path or his land.

Behind these images of "coal miner, once farmer," and the time frames they evoke is a story of active underdevelopment and exploitation in the central Appalachian coalfields, of which eastern Kentucky's coal counties are a part.* The transformation of a self-sufficient farm economy into an industrial economy has left succeeding generations of mountain people wholly dependent on external demands for the region's natural resources.

It is important to understand that not all of eastern Kentucky, or central Appalachia for that matter, should be treated as a coalfield. As Banks has pointed out, "It is very easy to overlook the diversity of the region and thereby make unfortunate generalizations. . . . Many eastern Kentucky counties, which were not totally dependent on coal operators for jobs, followed a path to development which was shaped, in large measure, by manufacturing interests" (1979a: 12, 21).

Because the decline in mountain agriculture accompanying industrial development was most pronounced in areas of coal and timber extraction, the present study is focused on those counties in which resource extraction was the bottom line of "development." Matters of time and space also require this restriction. In any critique of the role of the agricultural sciences in the industrialization of eastern Kentucky, those counties affected by manufacturing interests (such as Rowan and Carter Counties) must be included. Banks's research (1979a, 1979b) indicates similarities in patterns of agricultural decline in these and the coal/timber counties. Therefore, extension of this study to include the eastern Kentucky counties affected by the growth of manufacturing from 1880 to 1930 would provide a more comprehensive grasp of the relationship between the emergence of scientific agriculture and the decline of mountain farming. It is important to keep in mind that in the present study "eastern Kentucky" refers to the coal/timber counties where resource extraction determined the form of industrialization.

The Decline of the Agricultural Sector

John P. McConnell, President of the Virginia State Teachers College, described central Appalachia as found by the early white settlers as "one of the

*"Central Appalachia" refers to about 60 mountain counties in parts of eastern Kentucky, northeastern Tennessee, southwestern Virginia, and southern West Virginia. In 1968 the region had the most highly concentrated rural nonfarm population in the country according to the federal Appalachian Regional Commission (1968:1). "Appalachia" generally refers to a region of the United States along the Appalachian Mountain chain from southern New York to northern Alabama.

most favored sections of the Republic." Well suited to farming, the region was "unsurpassed anywhere in the number and variety of its natural resources: water power, a genial and salubrious climate, abundant but not excessive rainfall, good soil, and practically every fruit, grain, vegetable, or grass of the temperate zone" (1929:21).

By the mid-1880s an agricultural economy dominated most of the central Appalachian Mountains. Mountain farms of the period were well developed and self-sufficient, producing a wide variety of crops and livestock. Fifty years later, independent farming as a way of life had all but disappeared. Ron Eller has argued that "the traditional image of the pre-industrial mountain farm usually associated with the stereotyped picture of Appalachia was in fact a product of industrialization" (1976:5-6). Eller studied U.S. Census documents for 1880 through 1930 and noted a sharp decline in the average size of central Appalachian farms from about 187 acres in the 1880s to only 76 acres by 1930. This decline was most pronounced in the central coalfields and areas of intense economic growth. By the 1930s most mountain farms were only part-time units of production. The major source of income was from such nonagricultural jobs as mining, logging, and carpentry. Eller concluded that the expansion of industrialization into central Appalachia precipitated a serious decline in mountain agriculture.

Alan Banks (1979a:3-9) has examined census data for 1870-1930 for eastern Kentucky counties. The absolute number of acres in farms declined: "Bell County farm acreage, for example, decreased from 151,240 acres in 1880 to less than 60,000 in 1930. Harlan County farm acreage dropped from 220,000 to 57,850, and Letcher County figures fell from 255,500 to 135,446. In Perry County, farm acreage decreased from 237,456 to 135,446" (Banks, 1979a:7). The number of nonowner-operated farms increased (Table 2.1) as a pattern of sharecropping emerged, and the importance of the independent agricultural sector diminished.

Reliance on farming in central Appalachia persisted well after industrialization began. As late as 1925, Marshall Vaughn, the first editor of *Mountain Life and Work,* wrote that, of the 5.5 million people in the Appalachian Mountains, "six-sevenths of the people must make an independent living by working for themselves. That is done by tilling the soil of their own or rented farms. More than 4,000,000 people are farmers or make their living indirectly through the farmers by trading for them or selling goods to them" (1925a:4). Indeed, remnants of eastern Kentucky's farm economy were visible as late as 1933. C.J. Bradley, a researcher at the University of Kentucky Agricultural Experiment Station, noted that the "mountain counties constitute approximately one-third of the farms and almost one-third of the total land area of the state" (1933:155). By the 1930s, however, farm work was done primarily to supplement nonfarm family income.

As farming declined, a dependence on the growth of a new coal industry developed. In the early mining censuses between 1870 and 1890, southeastern

Kentucky coal production was categorized as "farmers' diggings" (Banks, 1979a:13-16). But, as Table 2.2 indicates, after the turn of the century the number of coal miners reflects the growing dependence on coal operations for employment. To further illustrate the importance which coal mining had assumed by 1930, Banks calculated the male mine employees as a percent of all male employed workers. He found that in 1930 coal miners made up 44 percent of the male labor force in Bell County, over 53 percent in Perry County, 65 percent in Letcher County, and over 75 percent in Harlan County.

County-by-county data are not available on the growth of employment in the timber industry. There is evidence, however, of a great surge in employment in commercial lumbering during the decades when farming was declining. According to Duerr, Roberts, and Gustafson (1946:9-12), 1880 marked a turning point in timber harvesting in eastern Kentucky. Before that time, timber clearing had been for agricultural purposes and primarily along bottom lands. Farmers cut trees to clear pastures and crop fields, to make tool handles and furniture, and to build barns, smokehouses, and other farm buildings. Around 1880, river logging developed into big business. Construction of the railroads gave a further boost to the timber industry, and the cutting of eastern Kentucky trees occured at a feverish pace until about 1923. "Few if any products of the forest were overlooked in this period of exploitation [as timber companies] from widely scattered points in the eastern United States and Canada invested in timber enterprises in the area"(Duerr et al., 1946:10).

Timbering in the Breathitt County area peaked within thirty years of the arrival of the railroad:

> Seven band mills, each cutting from 25,000 to 140,000 board feet per day, were built in Breathitt County, and two band mills in adjacent counties ran railroad spurs into the area. One of the four railroad-logging companies in the area built forty miles of main line track and ran spurs up hundreds of hollows. Stave mills and a number of large and small circular mills were in operation. Several planing mills and flooring mills, a dogwood shuttle-block factory, hickory dimension mills, and a chemical distillation plant were established. The Erie, Pennsylvania, New York Central and other railroads purchased thousands of hewn and sawn ties, and many merchants and lumber companies became tie dealers. Dealers in veneer logs, figure woods, burls, curly walnut, walnut stumps, and walnut logs traveled over the area . . . probably more than 25 million board feet of logs plus thousands of staves and ties were brought to or through Jackson annually on the largest [river] tide (Duerr et al., 1946:10).

At Quicksand, Kentucky, in Breathitt County, one of the largest of the hardwood sawmills employed 500 men to cut 140,000 board feet of timber daily (Duerr et al., 1946:12). Obviously, the number of laborers necessary to support such lumbering was substantial.

The dates of this Breathitt/Perry County area lumbering boom, 1880-1923, indicate that a shift to jobs in timbering occurred prior to the major coal

Table 2.1   Nonowner-Operated Farms as a Percentage of Total Farms for Selected East Kentucky Counties, 1880-1930

	Bell	Harlan	Letcher	Perry
1880	19.8	26.5	17.0	24.4
1890	24.3	32.6	27.95	24.4
1900	55.2	45.3	38.0	35.1
1910	55.0	58.13	40.0	39.1
1920	47.8	53.1	33.5	31.2
1930	47.3	44.0	34.0	36.0

Source: Alan J. Banks, "Labor and the Development of Industrial Capitalism in Eastern Kentucky, 1870-1930" (Ph.D. dissertation, McMaster University, 1980), p. 105.

Table 2.2   Number of Coal Mine Employees in Selected East Kentucky Counties, 1890-1930

	Bell	Harlan	Letcher	Perry
1890	0[a]	0	0	0
1900	304	0	0	0
1910	3197	0	0	0
1920	4079	7391	3989	4064
1930	3276	11920	5884	5088

Source: Alan J. Banks, "Labor and the Development of Industrial Capitalism in Eastern Kentucky, 1870-1930" (Ph.D. dissertation, McMaster University, 1980, p. 110.
[a] "0" indicates that no *industrial* production of coal was reported.

development and began to decline as the forests were exhausted and employment in mining moved to central importance. From 1880 to 1930, then, farmers who had been self-sufficient prior to coal and timber extraction were increasingly drawn into a new industrial labor force. The decline in the number of acres in farmland, the increase in nonowner-operated farming, the increase in the percentage of coal miners in the male labor force, and the rise of the commercial timber industry all indicate a serious decline in the agricultural sector in eastern Kentucky.

The Creation of a Resource Monopoly: A New Era Begins

Economic change in eastern Kentucky occurred as part of the expansion of capitalism. An entirely new way of life emerged in the mountains as farmers laid down their hoes and went from their pastures to the sawmills and the coal mines. But the changes did not come from "natural" forces or from the evolution of human societies. Rather, eastern Kentucky's transformation must be understood within the context of an expanding capitalist world economy.

Following Marx's perspective, Immanuel Wallerstein (1974; 1976) has shown that at one historical moment a singular economic transformation occurred and a "capitalist world-economy" developed in Europe (including Iberian America) between 1450 and 1640. Over time, that world economy has expanded outwardly, by incorporating ever more territorial units of the globe. It has also expanded intensively as processes internal to this economy worked themselves out, namely, the "proletarianization of labor" and "commercialization of land." Since the entire globe was incorporated into this new world system by about 1900, no areas can now be seen as autonomous. A capitalist mode of production depends on regions with nonproletarianized labor and noncommercialized land to maximize opportunities for overall profit. Land areas, then, are preserved for expanding production at low overhead costs. Similarly, pools of cheap labor are preserved. Further, in sectors where this occurs, the costs of labor are again reduced, since the local population bears the life-cycle costs of childhood and old-age maintenance of workers. David Walls (1976) has shown that an account of the worldwide capitalist economic system made of interdependent parts is needed to understand the central Appalachian region.

There is evidence of conscientious efforts to link eastern Kentucky with the expanding world capitalist economy during the aftermath of the Civil War. Following the Civil War, Kentucky businessmen and Bluegrass farmers worked through state politicians to develop policies to help them recover from the loss of slave labor. These policies were developed at a time when American and foreign capitalists needed timber, coal, other minerals, cheap labor, and new investment potentials (Duerr et al., 1946:9-16; U.S. Dept. of Agriculture, 1902:21-57; Gaventa, 1975:69-74). These policies endorsed by the state legislature helped trigger the economic exploitation of eastern Kentucky and led to the demise of its farm economy.

Banks (1979b:11-21) has shown that efforts to rebuild a labor system after the Civil War had great consequences for the mountain economy. Policies developed by 1875 opened the entire state to outside and foreign capital. The pattern of absentee ownership of land and minerals that predominates in eastern Kentucky today was thus established (Gaventa, 1973; Ridgeway, 1973; Eller, 1977; see specifically for eastern Kentucky: Millstone, 1972; Childers, 1977; Kirby, 1969). The post-Civil War policies also gave important support to railroad construction, including the first extension of rail lines into eastern Kentucky. Thus, railroads became agents for distributing both labor and capital throughout the state (Banks, 1979b: 19, 29).

Government programs were intentionally developed to attract white European laborers, to distribute them by rail across the state according to the needs of employers, and to attract outside capital. A Bureau of Immigration was established in 1871 with one agent in New York and two in Europe to contact chambers of commerce in European cities and to advertise investment potentials and job opportunities. Generous tax laws, the promise of adequate labor, and the lure of eastern Kentucky coal, timber, and iron were used to entice foreign capital (Banks, 1979b).

The first state geological survey of coal lands was authorized in 1854 and completed in 1859 (Turner, 1954:240), and other state surveys followed. Private surveys were also financed by railroad corporations and other industrialists. By the 1870s detailed estimates of timber, iron, and coal resources were readily available to investors. Samples of Kentucky coal, timber, and iron were even displayed in European industrial expositions in efforts to attract foreign capital (Banks, 1979b:16).

There is ample evidence of the success of this conscientious policy of advertising eastern Kentucky. Arnett (1978:32-40), for instance, reported that by 1885 an English company was looking at eastern Kentucky for potential investments and a Swiss company had purchased the Kentucky Union Railroad's Red River iron works, and by 1887 an English syndicate was buying land in Perry and Breathitt counties.

One of the best-known examples of foreign investment in eastern Kentucky during this period was the birth of Middlesboro in Bell County. In the four years between 1888 and 1892, some $20 million of British capital transformed the Yellow Creek Valley of Kentucky into "the magic city of the South." The investment included building railroads, furnaces, industries, hotels, streets, and schools. In short, an industrial city was built as the rich seams of coal were opened up to provide energy to such southern industrial centers as Birmingham, Chattanooga, and Atlanta (Gaventa, 1973:44-46 and 1975:42-110; Banks, 1979b:29-34).

By 1892 consolidation of land by absentee owners was clearly evident in eastern Kentucky's coal counties. Banks (1979b:23-29) has described the land investment of the late 1880s as "qualitatively" different from earlier, more casual land buying. First, the same names began to reappear in county after county on lists of the largest landowners. Next, a marked increase in absentee ownership became obvious. Banks concluded that a virtual land monopoly was established prior to the industrialization of the counties, with rail construction following monopolization of land holdings. This pattern of a series of big land deals followed by railroad building and then by commercial coal production has been also reported by Caudill (1962:65, 70-75) and Arnett (1978:41). The monopolization of land, encouraged by state officials and business interests, laid the groundwork for the economic growth that eventually destroyed eastern Kentucky's independent farm economy.

Mountain farm families felt the first tremors of the new order when land and mineral agents appeared in mountain hollows. According to Caudill, the first timber buyers came in 1875; and by 1885 the mineral agents—often with broad form deeds*—had followed (1962:72-82). Banks concluded from county tax records that 1880 was a pivotal year in land ownership changes (1979b:23-24). As Anna Rochester wrote in 1932:

> Big capital had taken possession in the wild mountain valleys of eastern Kentucky before 1914. During the war and since the war it has strengthened its hold over a steadily increasing area. Today Wall Street and capitalists in Chicago, Detroit, Baltimore and Cincinnati dominate the coal fields of Harlan and adjacent counties (1970:50).

She went on to document the arrival of the railroads and investment by such capitalists as Morgan, Insull, Mellon, Ford, Rockefeller, as well as such "native" investors as the Mahan family (1970:50-58).

A number of Appalachian scholars have stressed that these agents deceived farmers who were in no position to judge the value of their timber and minerals to industrial centers. In addition to having knowledge the farmers lacked, these agents often used trickery, terror tactics, and outright forgery to acquire property (Gunning, 1979; Wright, 1970; Gaventa, 1975:72-74; Caudill, 1962: 61-76). The difference between what farmers were paid and what their resources were worth proved to be startling. James Watt Raine, an early educator at Berea College, concluded that "the mountain people are suffering from the ruthless exploitation of large financial interests . . . foreign juggernauts [he argued, acquired the coal and timber] . . . for a song" (1924:236). Sara Ogan Gunning has recounted her own parents' loss of their land:

> My father and mother were born on farms in Clay County. There wasn't no coal mines then. They lived pretty good because they raised nearly everything they needed. What they didn't need, they'd take to the store and swap for what they

---

*The broad form deed gave the owners of mineral rights precedence over the owners of surface rights.

needed. That's how the coal men got the land—they bought some of the land, back in 1880, for $10 an acre and leased some of the mineral rights for 25¢ an acre. People didn't know coal was under the ground. Once they took the coal out from under the mountains, the land was so poor it wouldn't sprout seeds. The farmers thought that was so much money because they didn't know about money then. I'm sure if they'd have knowed it, they'd never have sold that land—it would have been their kids' birthright (1979).

Caudill (1962:81-82) has reported that a system of sharecropping developed from about 1900 to 1910. Coal and timber sales had put considerable amounts of money into circulation. Previously, the economy had been based on exchange of goods and small supplementary cash income from very restricted export trade in such products as salt, pig iron, agricultural surplus, timber, and coal (Banks, 1979b:5-11). With the increase in land and mineral sales, some county seat merchants and some farmers accumulated enough money to extend their landholdings. Frequently, the sharecropper was allowed to remain on what had been his family's land and deliver one-third to one-half of his corn and other crops to the landlord.

Banks (1979a:8-9) corroborated Caudill's description of this emergence of sharecropping with data from federal agricultural censuses. He found that the number of nonowner-operated farms increased from 1880 to 1930. Over three-fourths of the nonowners of this period were sharecroppers. A new class of local landlords was thus emerging on the heels of the creation of a resource monopoly by absentee financial interests.

The Building of the New Order

The reality that a new way of life was emerging first burst full force on the local scene in an almost frantic spree of building. The landscape of eastern Kentucky was literally carved into new railroad terminals, lumber mills, coal tipples, and industrial towns.

First, the railroad crews came. In some counties there was a ten-to-twenty-year lag between the big land deals and the extension of the railroads, but the construction of railroad links was well under way by the early 1900s. Caudill has described the building of the railroads as "the most momentous single occurrence in the history of the Cumberlands" (1962:75).

Many mountain people saw the coming of the railroads as a sure ticket to prosperity. One article in the *Jackson Times,* printed to attract small businesses to Breathitt County, is worth quoting at length:

The mountains of eastern Kentucky are at last to yield up the vast wealth of timber, coal, limestone and minerals, which so long have been neglected by the prospector, capitalist and manufacturer. The extension of the Lexington and Eastern Railway, now being made eastward toward the West Virginia line, the railroad extensions projected by the Chesapeake and Ohio Railway and other companies, and new roads proposed, all of which will afford an outlet for this untold wealth of raw material, and the feverish financial activity prevailing throughout this section of Kentucky is releasing among the people of eastern Kentucky hundreds of thousands of dollars each month. These facts tell the story of trade opportunities that wise men will not neglect . . . if you want to do business with the people of this section, you are earnestly requested to investigate these claims, and avail yourself of the splendid opportunities to increase your trade (1911).

Caudill has pointed out the irony behind such hopes that the railroads would bring prosperity to the mountain man: "Now the trees that shaded him were no longer his property, and he was little more than a trespasser upon the soil beneath his feet" (1962:76).

Existing towns were usually county seat towns, each with a courthouse and jail, a few small-scale merchants, sometimes a bank or wholesale business, and a number of homes. These towns, which had serviced an agricultural population, were inadequate for the resource exploitation that began in 1880 in Bell County and expanded in two decades to other parts of eastern Kentucky. Town building (usually) followed railroad construction. The county seats were so drastically changed that they seemed to be new, and many wholly new towns *were* built.

Lumber camps were built as the railroads insured for the first time a reliable means for moving timber out of the mountains. In 1910 and 1911 the Kentucky Hardwood Company "created" the community of Quicksand on land where Miles Bach had once farmed. The *Jackson Times** reported that the company was "erecting a class of buildings which are far ahead of anything heretofore attempted in this section. . . . Splendid entertainment is afforded at the company's clubhouse" (1911). When the L & N Railroad decided to put a terminal at Jackson instead of Lexington and a new hotel was built, the same newspaper labeled Quicksand "One of the Leading Little Cities of East Kentucky" (*Jackson Times*, 1911). A decade later 15,000 acres of virgin timber in the Quicksand area of Breathitt County had been reduced to lumber that was shipped to all parts of the world (Shackelford, 1979:9).

Coal camps, too, were being built. With amazing speed, construction crews erected small industrial cities where farmers had once raised corn and fattened hogs. In Letcher County 900 new houses and 8,000 people were soon living on "Bad" John Wright's cornfields and pasture lands at the new town of Jenkins.

---

*Two papers predated the *Jackson Times*. The first paper in the county, the *Jackson Hustler*, was started on December 28, 1888, by John Goff, Jr., of Clark County (Writers Program, 1941:88). The next July the Rev. J.J. Dickey took charge of the paper. He was a Methodist preacher from Flemingsburg, Kentucky, who also founded Lees Junior College in Jackson. The paper ceased publication in 1897. A second paper, the *Breathitt County News*, was started in 1901 and published until 1909 under J.W. Hagins, a Breathitt County native. Its circulation reached about 1200. Both papers reflected the same tone as the *Jackson Times*, announcing heartily that the county and Jackson had entered the ranks of progressive counties and that industrialization meant the arrival of the Promised Land.

Coal camp houses for miners' families were often built in terrace fashion, ascending mountain hillsides where virgin timber once stood.

This building of new communities was occurring all across the central Appalachian coalfields. According to Eller, most of these communities were built to facilitate expropriation of resources: "The majority of these new industrial communities were company towns. In fact, over six hundred company towns were constructed in the Southern mountains during this period, and in the coal fields they outnumbered independent incorporated towns more than five to one" (1976:6-7).

Larger coal companies such as Inland Steel, Consolidation Coal, U.S. Coal and Coke, Elkhorn Coal, and the American Association hired outside construction companies. Some skilled labor was imported from cities like Baltimore and Cincinnati (Caudill, 1962:98-99). Wages of $1.50 a day encouraged many mountain farmers to leave their fields and join construction crews. Cash came into many mountain homes where a cash economy had theretofore been very limited.

As Caudill (1962:99) has pointed out, the architecture of the coal towns varied. Homes built by heavily capitalized corporations were often well constructed with plastered walls, tar-paper roofs, front and back porches, and bright paint. Other companies, unable or unwilling to build more comfortable camp houses, threw up shoddy shacks. Some towns, such as Middlesboro, could boast resort hotels suitable for English royalty, administration halls, sidewalks, schools, and recreation facilities. Others consisted only of camp houses for miners, a better house for the mine manager, and a company store. Despite the variations, the rapid construction of the industrial towns marked a dramatic change in the lay of the land and the world that surrounded the mountain farmer.

A New Social Order Emerges

All the activity accompanying the creation of these new communities generated a sense of excitement and enthusiasm that spread across the eastern end of the state. As industrialization reshaped the physical world of mountain farmers, vast changes in social relationships and daily life occurred.

During the decades of transition, there was a startling population growth in eastern Kentucky. From 1870 to 1930 Bell County's population grew from 3,731 to over 38,000; Harlan County's population grew from 4,415 in 1870 to 64,557; in Letcher County population increased from 4,605 to 35,702; and the population of Perry County rose from 4, 274 to 42,186 (Banks, 1979a:5-6).

Many of the new residents in boom towns like Hazard, Quicksand, Jenkins, Whitesburg, Lynch, Harlan, Prestonsburg, Wheelwright, and Pikeville were laborers. First they built—carving mountain farmland and hillsides into the industrial structures of mines, mills, and company towns. Then they extracted—cutting the timber and mining the coal. These people put their muscles into the heart of the industrial development, bringing from the earth resources that promised great profits to others. These laborers were the numerical majority in the new social mix of people that was developing in eastern Kentucky.

Railroad construction brought an influx of European immigrants and blacks from the Deep South. As Caudill reported, these work teams generated much excitement: "People flocked from miles around to see the gangs of burly workmen driving steel and laying track" (1962:94). In Breathitt County the *Jackson Times* reported: "Gangs of laborers are pouring into eastern Kentucky to work on the extension beyond Jackson . . . a great deal of this labor is foreign. Altogether there are about 1,500 engaged on the extension, and work is being pushed with a vigor on the entire 80 mile division, extending from Jackson to Whitesburg" (1911). The Kentucky legislature's efforts to attract foreign labor as well as capital proved successful. Immigrants came from Italy, Hungary, Poland, Rumania, Albania, Greece, Germany, France, and Russia (Banks, 1979b:16-18; Caudill, 1962:103; Gaventa, 1975:77). Oral history abounds with stories of black laborers brought from the Deep South to work on rail crews who were brutally murdered by section bosses and left to die alongside the freshly laid track (Caudill, 1962:95; Worthington, 1978). Similar accounts tell of non-English-speaking workers from Europe who died building the new industrial order.

It is possible that the excitement of the construction and the influx of foreign and black workers led local newspapers and later chroniclers like Caudill to exaggerate their numbers in comparison with the great numbers of farmers who were drawn to the new industrial centers. It is also possible that, due to the brutality of construction section bosses and employers, many of the immigrants and blacks who came into the region were never counted in the population censuses. Many died on the jobs, and others moved constantly from one construction site to another as well as from county to county.

Still, it is important to recognize that farmers from the rural mountain regions of eastern Kentucky and central Appalachia made up most of the laborers, ditchdiggers, construction workers, and miners. Gaventa (1975:77) found farmers to be the majority among laborers in the Bell County development era. Banks (1979a:10-11) reported from population censuses for the years of eastern Kentucky's capitalist growth that "the native white population constituted around 90%, or greater, of the total." The Dreiser Committee, assembled in eastern Kentucky to investigate violence and terror used to block organizing efforts by coal miners, reported in 1932 that the great population increase in the Harlan coalfields came from mountain farmers (Levy, 1970:22). Finding it increasingly difficult to survive as sharecroppers, many mountain farmers left their farms to join the rail workers and construction workers laying track and building industrial structures alongside people from Europe and the Deep South.

Following on the heels of the railroad gangs, other work crews came for a second phase of building. Caudill described the speed of the construction: "In place after place the track layers had hardly passed from sight upstream when agents of the coal companies began the construction of coal camps, tipples and other installations" (1962:97). Again, native eastern Kentuckians joined with outside carpenters, plasterers, and masons who arrived for this construction.

Skilled artisans in eastern Kentucky often subcontracted with outside construction companies hired by coal corporations to build the camps. Isaac Jefferson Ward and his son Thadeus Stevens Ward, for example, subcontracted to build camp houses at such coal camps as Allais and Kodak in Perry County. A skilled carpenter and builder, Isaac Ward moved to Hazard from Magoffin County in 1892. He had been building in Magoffin, Breathitt, and Perry counties. He and his son built such historical buildings in Perry County as the 1912 Court House, the 1912 Hazard Missionary Baptist Church, the first school in Hazard, the second building that housed the Buckhorn Presbyterian Church, the Brown Johnson House at Chavies, and all the buildings of the Hazard Baptist Institute (Gabbard, 1979; Maggard, 1979).

Local merchants and professional people financed the building of some towns, usually located near corporate coal camps, as "satellite" towns. Many of these towns had a few businesses and a number of houses, often called "shotgun" houses, which were rented to miners or mill workers. Caudill (1962:108) described an "orgy of coal-town building" that continued until 1927 when the coal market collapsed. He listed 33 new towns in Letcher County by 1927, 37 in Perry County, 40 in Pike County, 25 in Harlan, and nearly as many in Bell, Johnson, and Floyd counties. Examples of these independent towns are Hi-Hat near Wheelwright, Neon and Blackey near Whitesburg, and Jeff near Hazard. Indeed, "between 1900 and 1930, the urban population of the [central Appalachian] region increased fourfold and the rural nonfarm population almost twofold, while the farm population itself increased by only 5%" (Eller, 1976:6-7).

A new group of merchants and businessmen emerged in the company towns and in new independent towns springing up around the industrial centers. These small entrepreneurs struggled to establish themselves in the new economy and in the new social hierarchy. Their economic success depended on the success of the coal and timber exploitation. As a result, these small businessmen, along with a growing number of professional people, welcomed agents from the financial centers of the nation and Europe with open arms. Front-page headlines in the *Jackson Times* (1911) announced tours by potential investors: "Louisville Boomers will be the guests of Jackson for parts of two days, a Royal Welcome will greet the Commercial Giants of Kentucky's Metropolis"; "Jackson is still on the map and the Louisville Commercial Club will honor the town by visiting on May 25"; and "Lexington and Jackson Exchange Greetings: Jackson is on the Map." The town was "honored" by a visit that lasted just two hours.

In an effort to prove that the town was a growing city and a likely place for profitable investments, the *Jackson Times* covered its front page several times in the spring of 1911 with a complete listing of all Jackson's businesses. The paper also gave prime news coverage to the opening of new businesses, announcing, for instance, in 1911 a new millinery, a new theater, a general store opened by a Winchester businessman who had moved to Jackson, and the joining of the New York Fidelity and Casualty Company with the local Jackson Real Estate and Insurance Agency. Other news stories covered meetings of a newly organized "Women's Auxiliary" of the local Commercial Club, with references to the similarities between Jackson's club and women's clubs in Lexington and Louisville.

It is clear that a sizable part of eastern Kentucky's preindustrial elite was particularly eager to be recognized as social equals by the fleet of industrial administrators and professional people who accompanied the openings of the large industrial centers. Rylan C. Musick, editor of the *Jackson Times,* reflected this attitude, as references to his newspaper have already shown. Musick's paper started publication in late 1910 and is one of a number of eastern Kentucky county newspapers that were organized by local business-men at the time of industrialization (Maggard, 1977). This paper is cited heavily here because it is the county paper located where the University of Kentucky built its branch Agricultural Experiment Station, which was to serve eastern Kentucky farmers. References from this paper help to establish the climate in which the early extension workers forged policies of eastern Kentucky farm support. Its early years reveal an attitude of enthusiasm and endorsement of industrialization as experienced in eastern Kentucky. The same legitimizing spirit fills the early years of other county papers such as the *Floyd County Times,* the *Middlesboro Daily News,* the *Harlan Daily Enterprise,* the *Hazard Herald,* and the *Pike County News.*

A series of articles appearing in the Jackson paper before and after a visit to Jackson in May 1911 by Louisville and Lexington investors reveals frustra-tion and anger by the local business establishment over affronts and insults dished out by the outside capitalists. One headline announced: "Citizens Indignant" that this trainload of Louisville businessmen might skip Jackson on its eastern Kentucky tour. The Jackson Commercial Club rose to the occasion and assured central Kentuckians that there was, in fact, a decent place to spend the night, that they would not have to go armed, and that an undertaker need not bring caskets. Musick wrote an angry editorial pointing out that "Louisville is receiving hundreds of thousands of dollars from our people each year!" (the *Jackson Times,* 1911).

Some merchants in the new towns were from cities in the South or the East, hoping to make a new start in the boom economy. This also appears to have been a motive among some professionals who came during these years. Corporations also brought in doctors, nurses, teachers, lawyers, and preach-ers to "staff" company towns, which were built at so rapid a pace that the local elite were too few in number to meet the need. In addition, a cadre of missionaries from nearly every major church denomination in the nation was arriving in eastern Kentucky, frequently coming just as the railroad extensions were completed.

Even though many people moved into eastern Kentucky from distant cities during these years, it is incorrect to assert, as Harry Caudill has, that "the officials of the coal companies found little of value to them except the coal itself and the mining timbers growing on the earth above it" (1962:97). Caudill has argued that the whole network of people needed to industrialize the region was imported. The Perry County Ward family is typical of the preindustrial elite that moved into positions of leadership in the new era. Isaac Jefferson Ward and his son Thadeus Stevens Ward were not only builders and contractors. They were also teachers. Two more sons of Isaac Ward, William Thomas Ward and James Augustus Ward, were skilled surveyors/engineers. They had been educated at the National Normal University of Lebanon, Ohio. Two more sons were lawyers, John Downing Ward, educated in Nashville, and Samuel Melton Ward, educated in Louisville, who later served as circuit judge. John Ward also served as a secretary of the state board of education. To finance their education, children from such families would alternate periods of training outside of the region with periods of work at home (Gabbard, 1979; Maggard, 1979; Pilcher, 1913).

At every point, mountain people joined with new people coming in as teachers, merchants, engineers, builders, coal miners, preachers, business managers, and even coal operators and timber-mill owners (Pilcher, 1913; Gabbard, 1979; Campbell, 1969; Arnett, 1978).

At the top of the newly emerging social hierarchy was an upper class of absentee owners who were either European capitalists or Americans from wealthy elites of northern cities. Occasionally these people visited the region, but their presence was more often felt than seen. Their local representatives were the superintendents and managers of their investments, usually men who had experience in British mining or in older mine fields like those in Pennsylvania and West Virginia. Similarly, timber-mill superintendents moved across the country as lumber corporations cut through forests in a line from the northeastern forests, through the Appalachians, and into the South and Southwest.

Social distance between the superintendent elite, professionals, merchants, and laborers was physically evident for everyone to see day after day. Social distinctions were literally built into the new towns. European immigrants were often housed in "hunky-towns," or special neighborhoods in coal camps. Residents of Lynch, for instance, today can point out the "foreign" boardinghouse. Black miners were severely segregated. Benham was built next door to Lynch by a well-capitalized corporation to house black families. It had its own school and recreation facilities. Usually, however, black miners lived in miserable housing. Supervisors, on the other hand, lived on "silk stocking row," a series of better-built houses usually higher up the mountainside and overlooking the rest of the coal camp. Professional people could often rent homes in this elite section. It was with this group, living higher on the mountain, and with absentee investors, that local merchants and professionals sought acceptance. Professional people formed chapters of national associations in order to emphasize their legitimacy. A Mountain Press Association, for instance, was organized in 1911 and tried to convince the Kentucky Press Association to hold a meeting in Jackson.

Along with the spatial outline of social distinctions, place names helped reinforce the social hierarchy. Some camps, such as Haymond, Fleming, Jenkins, McRoberts, and Lynch, were named for corporate executives. Others were named for wives or women friends of executives—Bellcraft, Dalna. Others were named for the initials of coal corporations—Vicco and Seco. Street names in Middlesboro reflected places frequented by the British industrial elite who built the town. Names drawn from nature or local history were kept only for the work places and areas near the mines.

Some of the more prosperous farmers, merchants, and professional mountain residents formed lucrative partnerships with outside investors and joined the industrial elite. John C.C. Mayo of Paintsville, the son of a Johnson County farmer, is one of the most famous. He teamed up with P.L. Kimball of Chicago and Frank Buell of Sharon, Pennsylvania, to form Northern Coal and Coke, which led to larger operations in Letcher, Floyd, and Pike counties. Mayo later sold to Consolidation Coal, with the Mayo family remaining a major stockholder. When Mayo died in 1914 he and his associates owned 500,000 acres of land, and he left a $20 million estate (Arnett, 1978:40-42).

An Ethos of Progress

An ethos of progress through industrialization grew up rapidly. It was particularly strong among ambitious merchants and professional people whose profits hinged on the success of industrial expansion and whose prestige in the new social order depended on recognition by the industrial elite. Two editorials in the *Jackson Times* (1911) illustrate this sentiment. On March 17 the editor wrote:

> Movements of prospecting parties in the mountains near here are causing some considerable surmising among the people of Jackson this week. There need be no surprise manifested, for the wealth stored away in these hills is commonly known and it is but natural to assume that enterprising capitalists, sooner or later, would be led to develop our resources. Jackson citizens may some day wake up to a peaceful revolution, upsetting and re-arranging conditions and making for the betterment of affairs generally. Are you ready to receive the benefits of the change, or are you one of the complacent ones, without energy or initiative, who cares not for better schools, better roads and more modern and comfortable conditions?

A second editorial on March 31 blatantly denounced the preindustrial culture and history:

> The time will soon arrive when the man from the outside will come among you to reap where you have sown. Why not be forehanded and place yourself at the head of the march as a progressive citizen? The past can do nothing for you. The future holds much of good in store.

The local media in Middlesboro also applauded the role of absentee industrialists for their help in "developing" the Clear Fork Valley in Bell County. *The Middlesboro Daily News* (1965) published an "Anniversary Special Edition" in 1965 and claimed that the town was indebted to the American Association, Inc. Gaventa (1975:63) pointed out the irony in this loyalty. The company failed despite spending over $20 million to develop mines, railroads, a steel plant, industries, hotels, and a town on its 80,000 acres of coal and land. The local economy collapsed, and thousands of people lost their work and homes. The "development" had been entirely around the interests of the export-oriented industrial elite—to provide iron, coal, and steel for the rapidly industrializing South—with no concern for any future for the locality independent of the company.

Theodore Dreiser has written of the allegiance to the new order among much of the population:

> The small town bankers, grocers, editors and lawyers, the police, the sheriff, if not the government, were all apparently subservient to the money and corporate masters of their area. It was their compulsion, if possibly not always their desire, to stand well with those who had the power to cause them material or personal difficulties and, against those, the underpaid and even starving workers, who could do nothing for themselves (1970:10).

Some of the merchants and professionals owed their jobs and their rented homes to the corporations. Allegiance to the industrial elite among these groups did not arise solely from faith in "progress" as defined by industrial development. Rather, corporations were able to maintain tight control over employees—administrators, professionals, and managers of company stores as well as coal miners and mill hands. The British interests backing the Middlesboro development, for instance, retained controlling interest in most of the area's development. They owned the land and minerals and would only lease to other firms; they owned the railroads and charged royalties for haulage fees; they controlled the company that built the town as well as the corporation that built the Cumberland Gap Park as a luxury resort (Banks, 1979b:29-34; Gaventa, 1975:78-79).

Caudill (1962:100, 113-15, 174-76) has shown that many of the larger companies dominated every phase of life in their towns. Schools, churches, hospitals, and company store managers were all instruments of the owners and reflected their philosophy of management and business, as they were indebted to them for jobs. To get and keep jobs, miners were often required to live in the towns and in company-owned houses, to shop at the company commissary, to go to company doctors, to send their children to company schools, and to attend churches where they heard company-hired preachers (see also Levy, 1970:26-31, and Alinsky, 1970). Moreover, Pope (1942) has noted the importance of preachers in spreading the doctrine that the good of the community was directly tied to the success of the industrial developers.

Although there were differences in the nature of corporate control over life in the coal camps, it was always clear that the dominant influence was the coal owner and the mine bosses. Miners and their families, merchants, professionals, and administrators were always aware of the source of power in these coal towns.

SCIENTIFIC AGRICULTURE AND MOUNTAIN FARMING

The energy and timber corporations that have poured millions of dollars into the "development" of central Appalachia have realized great profits at the expense of local communities. In 1967 almost twice as much capital flowed out of the region as into it, entering the financial markets centered around New York (Checchi and Company, cited in Gaventa, 1973:43). The early pattern of absentee ownership and profit continues, and monopolization over the region's resources also continues (Ridgeway, 1973). In 1979 the Eastern Kentucky Housing Task Group reported to local, state, and federal agencies that families seeking decent housing were blocked by "the stranglehold on land availability by out-of-state and foreign corporations" (O'Donnell, 1979:1). *Dun's Review of Modern Industry* reported that "for all their small numbers . . . these coal royalists hold what may well be one of the most lucrative investments in all of America" (Murray, 1965:40). Rather than a future of prosperity, eastern Kentucky families have suffered intermittent crises of desperate unemployment, poverty, pitiful health care, undernourishment, substandard housing, poor schools, and other ills (Weller, 1972:47-55; Appalachian Regional Commission, 1972, 1975).

The College of Agriculture at the University of Kentucky, the Kentucky Agricultural Experiment Station (KAES), and the Cooperative Extension Service (CES) did not oppose this exploitation. Instead, while they were establishing their programs as legitimate within the university and with the state legislature, they developed policies and an orientation toward farming that tended to support the economic and social exploitation of eastern Kentucky. The College of Agriculture, KAES, and CES came to define mountain farming as supplementary to employment in mining or the timber industry. Farming along mountain valleys and on hillsides became defined as "subsistence" in comparison with the large flat farms in Kentucky's Bluegrass region. Loss of the land base crucial to mountain farmers was treated as part of the cost of "progress" in coal counties (see Craig, 1932:22-36). Agricultural scientists played a "vacuum-cleaner" role in eastern Kentucky. They researched and taught ways to restore hillsides ravaged by timbering and coal mining, to purify polluted streams and to increase family incomes in an industrial economy that fluctuated wildly with demands for resources. Since farming was only an adjunct to coal and timber employment, no serious efforts were made to develop a technology or science suited to mountain farm problems.

A look at the economic and social realities facing Kentucky's earliest university farm specialists suggests that it is hardly surprising that mountain farming was reduced to a category of subsistence and supplementary activity. Advocacy of the vanishing and powerless mountain farmers would have pitted the university against the political leaders of the state who backed and helped initiate eastern Kentucky industrialization. This stance would have blocked chances of getting the emerging College of Agriculture recognized as legitimately worthy of state funding. In addition, a national orientation to scientific farming was emerging in which mountain farming could only be seen as "marginal." University-trained agricultural scientists were developing a science which served the larger, prosperous farmers attempting to compete in a world agricultural market (Hightower, 1973; Rosenberg, 1976; McConnell, 1969; Hadwiger, 1975). To these scientists, progressive farmers seemed to be those on large land holdings in areas where the flat terrain was suitable for mechanization and high-productivity farm programs—an image that mountain farms did not seem to fit. Scientific agriculture, as it was emerging in Kentucky and the nation in the late 1800s and early 1900s, excluded regions like eastern Kentucky—no matter how developed their farm economy had been. Criticisms were occasionally raised, but they were not powerful enough to overcome the emerging new industrial order in eastern Kentucky and the shifting national orientation to farming.

The Science of Farming: Getting Started at the University

The state of Kentucky lagged several years behind other states in establishing its College of Agriculture, although the Morrill Act of 1862 encouraged the state legislature to participate in the federal granting of public lands (Smith, 1965:4). The Kentucky legislature accepted the federal land grant of 330,000 acres on January 27, 1863, and announced open bidding from colleges in the state for the agricultural and mechanical arts program called for in the Morrill Act. It was the intention of the Act that states would sell the land as a sort of land scrip and use the interest on the money to endow and maintain an agricultural program. However, the Kentucky legislature, apparently unhappy with the single bid it received from Transylvania University, lost over a year's time in establishing the endowment. The exact nature of the political controversy surrounding the location of the new program remains unclear. The Board of the State Agricultural Society, an organization officially recognized by legislative act and approved by Governor Morehead on March 10, 1856 (Turner, 1954:235), had accepted Transylvania's bid in December of 1863. The legislature, however, voted to reject the bid and hold bidding open another year (Bowman, 1868, 1872). Eventually a compromise was worked out through a merger between Transylvania and Kentucky universities. On February 22, 1865, Kentucky's Agricultural and Mechanical College was created as a department of Kentucky University, a "Christian Church institution reorganized by consolidation with Transylvania University" (Smith, 1965:4). Eventually the land scrip was invested to return $9,900 annually for Kentucky's program, an amount considerably less than several other states had managed to establish and one that hampered early efforts to develop the program (Bowman, 1868, 1872). The first students were admitted in October 1866 for apprentice-type practical training on a 433-acre farm set up on the recently purchased Henry Clay Estate and the adjoining Woodland Estate.

Once established, the first university program clearly included a commitment to aid mountain farmers. John B. Bowman, the Regent of Kentucky University, was enthusiastic about the new Agricultural and Mechanical College. He set about designing the program with a liberal and democratic flair in keeping with the overt intent of the Morrill Act. The wording of that Act implied a widespread national interest in practical education for farmers that would help establish a "permanent agriculture and stabilize farming populations" (American Association of Land-Grant Colleges and State Universities, 1962:2-5; Berry, 1977:147). Bowman saw Kentucky's program as especially important for poor students, and he included mountain students in his vision. Consider his 1872 appeal for funding from the state:

> I come before you now, gentlemen of the Legislature, not on my own behalf, but on behalf of the poor young men of our commonwealth; the men of the mountains as well as those of the plain . . . . I want those who live fifty or one hundred years hence to see in the Mountains of Kentucky the sons and grandsons of those whose intellects were first awakened and whose genius was evoked by the men who now labor to give the state an Agricultural College worthy of her (Bowman, 1872:15-16).

Bowman (1868:17-18) designed a three-part program for the farm. Students would learn practical farming on the estate's pastures, tillable lands, cattle lots, vegetable and fruit gardens, dairy, grass, and grain fields, and flocks and herds. A professor of horticulture and landscape would train them on the estate's ornamental grounds, experimental plots, orchards, nurseries, viticulture, propagating houses, greenhouses, arboreta, botanical garden, nursery, and seed business. In a mechanical shop, students would learn carpentry, wagon-making, blacksmithing, tool-making, and other skills needed to maintain a farm. While some of the existing faculty were able to initiate parts of these plans, funds were insufficient to hire the principal agricultural faculty Bowman needed.

Disputes with more conservative opponents in the Christian Church added to Bowman's problems of inadequate funding. By 1878, the legislature withdrew its support for Bowman and placed the college under full state control at the University of Kentucky (Smith, 1965:4).

The first "Course of Instruction" in agriculture and horticulture under the new administration was listed in 1878, but no professor was listed until 1880. President Patterson hailed the appointment of Professor W.A. Kellerman as head of the department of botany, agriculture, and horticulture: "For the first time since the establishment of the College a decided step has been taken in the direction of realizing the Congressional idea of instruction (theoretical and practical) in scientific agriculture" (Smith, 1965:5). Kellerman's appointment initiated the experimentation and research that gave shape to a curriculum and laid the foundation for a "new" agriculture. It had taken a decade and a half since the initial land grant for Kentucky to initiate courses of instruction and research. The next five years were spent developing that fairly limited program.

President Patterson needed more money to expand the program, and he faced the problem of selling the new program to the state legislature. By the 1880's many Kentucky farmers were dependent on tobacco production and were facing serious economic problems from overproduction, price instability, and control over the market by the American Tobacco Company monopoly. It was the larger tobacco farmers in central Kentucky who wielded considerable power in the state and whom Patterson had to sway. John Bowman's speeches on behalf of poor farmers and mountain youth would have had little persuasive power with central Kentucky tobacco farmers fighting to survive against the Duke monopoly over the nation's tobacco "industry." These farmers wanted to hear how a better-funded university program could help them fight the Tobacco Trust.

Patterson turned to other states for examples of how to gain the confidence of those farmers who would be allies in Frankfort. By May 1885 Patterson was convinced that experiment stations would greatly improve relations with powerful farm interests. He had attended a meeting in Washington, D.C., with representatives of other colleges and was so impressed with their work that he toured Wisconsin, Michigan, and Illinois to observe experiment stations already in operation and to seek a director for Kentucky (Smith, 1965:6-10). In September 1885, two years before the Hatch Act provided federal money for experiment stations, Kentucky established an Agricultural Experiment Station as part of its Agricultural and Mechanical College. The KAES responded immediately to Kentucky's larger tobacco farmers. Beginning with its first research bulletin, the KAES reported research for over three decades aimed at helping tobacco farmers improve their competitive position—research that tended to benefit the more profitable and highly capitalized farmers who were in a position to implement the new techniques that the university researchers recommended. Eastern Kentucky farmers rarely figured in these early years of experimentation and research.

The first KAES director was 25-year-old Professor Melville Scovell. He had been a special agent for the United States Department of Agriculture in charge of phases of sugar production in Kansas and Louisiana. Professor George Roberts, head of the Department of Agronomy of the KAES for many years, said that Scovell "knew that the future lay only in one course, that of immediate service to agriculture that could be readily utilized and appreciated by the average farmer of the State" (Smith, 1965:7). But, again, the title of the first bulletin distributed by the KAES indicates that Scovell and his staff perceived "average" farmers to be those tackling the problem of increasing tobacco yields: "Do Fertilizers Affect the Quality of Tobacco?"

As "practical and useful" information began to come from the KAES, the legislature responded favorably. In the spring of 1886 the state gave 12 acres of tillable (although quite poor) land to the KAES for field experiments. The 1887 Hatch Act assured Scovell of an annual grant of $15,000 from the federal government for research and the dissemination of research findings through free bulletins. By 1890, KAES owned a 48½-acre farm, had let the contract for a building, had added new staff, and had published 31 bulletins distributing 10,000 copies of each. In 1892, Scovell bought two cows, the beginning of the University's dairy herd.

The Agricultural and Mechanical College changed directors five times after Kellerman left in 1885. Clarence Matthews, the last of the five, headed the instruction program for the next 20 years. Enrollment increased and the curriculum was settled by 1907 with degree courses in horticulture, animal husbandry, and general agriculture. A two-year course of practical training was initiated, and a 10-week winter short course was offered. Special programs for women were taking shape, growing from a domestic science class held in the basement of the women's dormitory in 1905 to a Home Economics Department in 1910.

In 1908 the state legislature created a new "College of Agriculture" as one of several colleges in a "State University, Lexington, Kentucky" (Smith, 1965:12). Patterson, who felt that KAES had become a self-contained entity, urged unification of the program. In 1910 Scovell became Dean and Director of the College of Agriculture and KAES, and he requested state money to enlarge the program. Scovell, as Patterson had done before him, used other land-grant programs as yardsticks, particularly those at Illinois, Missouri, and Wisconsin. "We could compete with any of the agricultural colleges and build up, I believe, a great institution," Scovell told the legislature in 1910 (Smith, 1965:12). It was clear that the university's agricultural scientists felt keen competition within the academic world. Not only did they want to impress leading farmers and the state legislature with their worth, they also sought professional acceptance among their peers in the newly emerging science of agriculture. In 1911, the College of Agriculture had its first building, with rooms for agricultural machinery, offices, classes, a museum and assembly area, dairy, botany, agronomy, and soil laboratories, a greenhouse, and photographic darkroom.

The early 1900s were years of development for the KAES. Boosted by federal feedstuffs laws, food and drug laws, and amended seed laws, the KAES's inspection and enforcement duties had expanded beyond its earlier requirement to analyze all fertilizer sold in the state. According to J. Allen Smith, "interest of the farm people in the work of the Experiment Station and growing confidence in the value of research results led at last, in 1912, to an annual appropriation of $50,000 for the work of the Station" (1965:14).

By 1912 the College of Agriculture and the KAES were unified in funds, buildings, land, equipment, and staff. The farm had grown to 243 acres. The KAES staff had 42 members, and the college had 11 teachers. Enrollment had risen from 10 in 1903 to 101 in 1910.

As KAES research work progressed in the early 1900s, a body of scientific findings built up that researchers needed to disseminate to farmers. The KAES staff, however, wanted other people to get the word out, because extension work per se would interrupt their research. This sentiment existed in many universities and helped create a divison of labor that tended to insulate researchers from the farming population. In Kentucky the division between research and extension personnel meant that the special needs of mountain farmers became even further removed from the focus of university agricultural research.

Well before the federal Smith-Lever Act of 1914 was passed providing funds for extension programs, Kentucky was claiming extensive "contacts with the public" (Smith, 1965:17-18). Tommy R. Bryant was assigned to head all extension-type activities under the KAES in 1910. This work included the organization of boy's corn clubs to distribute seeds and bulletins, "movable schools" in home economics, an Agricultural Train with exhibits and lecturers in cooperation with the railroads, and the first "Farmers' Week" in 1911. In 1910 breeding associations and the Kentucky Corn Growers Association were invited to hold annual meetings at the College of Agriculture. In 1912, two extension field agents for statewide duty were hired. The first county extension agents were appointed for "field studies and farm demonstrations" in cooperation with the Office of Farm Management of the USDA, with counties paying half the costs. Early in 1914, before the Smith-Lever Act, 17 women were hired for short-term summer instruction in canning.

Although the Kentucky legislature adjourned before the Smith-Lever Act was passed, the president of the university and the Secretary of Agriculture signed a memorandum of understanding so that Kentucky could receive federal support. An administrative division of the College of Agriculture was set up to administer all extension work by the university and by the USDA in the state. In March 1916 the Kentucky General Assembly allocated $18,000 annually to match Smith-Lever funds and authorized county fiscal courts to appropriate funds as desired for extension workers. In 1918, after considerable confusion and feuding among what were now three branches of the university's agricultural work, Thomas Poe Cooper was named dean of the college, director of the KAES, and director of the Extension Service, uniting all three responsibilities. By 1919 there were 82 professional and 30 clerical workers in Lexington, 64 county agents in 60 counties, 36 home demonstration agents in 25 counties, 68 courses in resident instruction in agriculture and 11 in home economics, and 253 students enrolled. J. Allen Smith concluded that by this time "the College faced a long period of continuity in policy and practice, an opportunity to grow and expand in accordance with a continuing philosophy of education, research, service, and administration" (1965:22). Unfortunately, the desire to serve mountain farmers so clear in John Bowman's vision some fifty years earlier was lost as this "policy and practice" had evolved.

Eastern Kentucky Farming: A Nonissue

From the earliest days of organized agricultural activities in the state, eastern Kentucky farmers were not included. A Kentucky Agricultural Society was organized in 1837 with branches affiliated across the state—except in eastern Kentucky. The Society met primarily to exchange information on crops and farming methods (Hopkins, 1938). In 1856, when the state legislature officially recognized the organization, the Kentucky State Agricultural Society divided the state into three "agricultural districts." Eastern Kentucky was not included; instead, it was labeled the "mountainous region" (Turner, 1954:252). This occurred despite the fact that many of the same crops and livestock were raised in eastern Kentucky as in the rest of the state: corn, hogs, hemp, dairy and beef cattle, poultry, wheat, rye, sorghum, flax, potatoes, peas, maple sugar, beans, beeswax, honey, oats, and sheep (Verhoeff, 1917:138-42).

Early efforts to get the state to support geological surveys of coal resources had heavy support from agricultural societies. According to Wallace Turner (1954:240-52), a number of societies signed a petition to the Kentucky legislature in 1853 to support a geological survey. The first survey was authorized the next year. After five years the "Owen Survey" reported on mineral wealth in the "Great Appalachian Coal Field" which included all of 15 eastern Kentucky counties and much of five others. These societies saw this state geological survey as the beginning of a "scientific agriculture," and the Owens team felt it could help bring "practical results in the agricultural sections" (Turner, 1954:251-53). Eastern Kentucky, however, was not considered one of those sections.

Early organized farm activity was based in central Kentucky. Brutus Clay, of the powerful bluegrass region Clay dynasty, was the first president of the state-recognized Agricultural Society. The first state fair was held at the Bourbon Fair Grounds near Paris on September 30, 1856. The Society, under Clay's leadership, sponsored field trials of new machinery such as reapers and mowers near Louisville in 1858 and near Lexington in 1859, encouraging the new farm-implement businesses to focus on Bluegrass farm mechanization potential. The Kentucky Harvester Company awarded prizes at the trials.

It was the organized farm population engaged in these activities that raised the demands for a state agricultural college by the late 1850s (Turner, 1954:230-40)—people who had not even defined eastern Kentucky as an agricultural district. It is clear that, long before the first students knocked on the doors of Kentucky's first Agricultural and Mechanical College, the leading figures of the state's farm population had already discounted eastern Kentucky farmers. Therefore, it is not surprising that the emerging university farm programs did not devote staff and research resources to develop a special hillside technology suited to mountain climate, terrain, and soil.

The 1912 state appropriation of $50,000 annually to the KAES provides a case in point. Recollections by George Roberts, head of the Agronomy Department, indicate the intensity of the pressure he and his staff felt in 1912 when the funding was announced. The money was intended in part for fields for soil experiments. Roberts wrote that he had to choose field sites quickly "for it was felt to be highly important to get results as quickly as possible to justify the appropriation" (1944:57). No experimental soil plots were selected in eastern Kentucky. According to Roberts, "The location of the fields was highly important because they should be on typical soils of large farming areas and should be so located as to be accessible to farmers for meetings for demonstrations and educational purposes" (1944:57). Selection of a plot in eastern Kentucky that would be "accessible" to mountain farmers for their education and "typical" of their soils was not even a consideration. Such a move would not have helped "justify the appropriation," the key motivation in site selection.

Much of the research at the KAES in the early 1900s made it appear as if all Kentucky farmers lived on identical farms located in identical climates and had the same needs. A review of KAES bulletins sent to newspapers across the state from 1911 through 1915 reveals articles that could have been of particular interest to mountain farmers, but no focus on regional differences appears. The same advice was handed out to all farmers, despite the obvious fact that climate, credit availability, terrain, and soils were quite different in the mountains.

Consider bulletins on orchard care as examples. L.R. Neel, Associate Editor of the *Southern Agriculturist,* wrote in 1927 that for mountain farmers, a special agriculture exists in which "the tractor and large tillage tools are to a great extent precluded . . . yet agriculture has an important place in the southern mountain region" (1927:4). He cited orchards as an important part of mountain farming. Professor Harvey S. Murdock, President of Witherspoon College in Perry County, recognized the importance of orchards in eastern Kentucky farming. In 1922 Murdock hired E.C. Jones, who had 25 years of experience in orchard care, to work with his school in developing varieties of fruit trees adapted to the mountain soil and climate (*Jackson Times,* 1923:7). Floyd Bralliar in 1927 pointed out that for many years the mountains had been ideal for fruit trees: "They will thrive and bear good crops on the naturally acid soil of the mountains, the high altitude and the natural drainage of both water and cold air combine to fit the mountain country peculiarly for fruit growing. . . . Fruit growing offers one of the greatest agricultural opportunities to the southern mountain farmer" (1927:17, 25). Yet bulletins from the KAES to newspapers, including those in eastern Kentucky papers, sounded as if the conditions for raising fruit in the Bluegrass region were universal throughout the state. The same was true for bulletins concerning soil improvement, dairying, corn crops, insect control, weed control, forage plants, poultry, grapes, and sheep (KAES, 1911 through 1915).

Research program directors appear to have been responding to the "squeaky wheels" of powerful central Kentucky farm interests who could control their money and program expansion. As a result, extension workers in the eastern end of the state were carrying a body of research findings to mountain farmers which frequently had little relevance to their problems. But impoverished and politically unorganized mountain farmers could not help insure a well-endowed university program with plenty of money, staff, and laboratory facilities for research. These latter considerations were paramount among agricultural scientists who wanted to win professional recognition among their academic peers at other land-grant colleges. According to Rosenberg, at most of the land-grant colleges, making allies with powerful farming interests was a logical step since it promised "the opportunity to perform the research and write the books and articles which alone could gain for the agricultural scientist the recognition of his disciplinary peers. And this recognition was, for many, the only sort of success that ultimately mattered" (1976:179). A selection from Rosenberg's *No Other Gods* is worth quoting at length for its vivid description of the world in which early American agricultural scientists negotiated their programs:

> Agricultural opinion is not made by some undifferentiated group of hard-working farm operators. It is made by the articulate, the prosperous, and the influential both within and without the community of agricultural producers. Educated and more highly capitalized farmers, editors of farm and rural papers, country bankers, insurance agents, merchants, and implement dealers make up the visible agricultural consensus. These were the men active in most farmers' organizations and specialized producers' associations—those who had come to accept scientific knowledge as necessary for successful economic competition. Their letters filled the correspondence files of experiment station directors and agricultural college deans. Naturally, these men of consequence met station workers, deans, and directors at meetings, at farmers' institutes, and at social gatherings. Editors came habitually to call upon station directors for columns of free copy, while politicians turned to the same men for sympathetic and knowledgeable speeches on agricultural questions. It was inevitable that experiment station workers should play increasingly important roles in the affairs of farm organizations; it was equally inevitable that these farm associations and the rural lawyers and editors connected with them should become the backbone of college and station support in the state legislatures (1976:176-77).

In eastern Kentucky, farmers were faced with a loss of control over the land base, increasing abuse of land by coal and timber companies, and a frantic burst of industrial development during the years that powerful farm interests in the rest of the state were shaping the College of Agriculture. It would have taken unusual, even rebellious people at the University to step beyond the social, economic, and disciplinary pressures on their work and respond to the needs of eastern Kentucky farmers. McConnell has pointed out that the inclinations of people attracted to the agricultural sciences made such a stance highly unlikely in any land-grant college. Early agricultural scientists were "oriented toward natural science by training and by proclivity. They regarded themselves as nonpolitical and it may be hazarded that as a group they strongly partook of the ideological suppositions of the existing economic order" (1969:23). The "existing economic order" in eastern Kentucky was hardly one that supported expanding a mountain farm economy. A comment by Bryant in a 1911 KAES bulletin indicated that the vision of the first director of extension at the University of Kentucky was, in fact, supportive of the new economic order in the mountains: "Railroad officials are businessmen and see things from a business standpoint, and the farmer and school teacher must realize that their success lies in their taking a similar view of all affairs of life" (1911). Taking a "business standpoint" in eastern Kentucky, as the first section of this chapter shows, meant closing the door on an independent farm population and plunging farm families into an unstable and vulnerable industrial labor force.

McConnell has pointed out (1969:28) that before 1900 the USDA was not a force of consequence in university agricultural programs. The Hatch Act had raised some fears of federal control among agricultural scientists, but it was the passage of the Smith-Lever Act in 1914 that highlighted growing federal supervision over state university work. To get federal allocations for cooperative extension work, colleges had to agree to reorganize existing extension programs, spend not less than 75 percent of the money on extension, and submit to supervision by the USDA (McConnell, 1969:35). From the beginning, business interests in positions to benefit from farming supported efforts to establish a national extension service. McConnell (1969:20-35) has described the business firms and associations active in backing the new farm "demonstration" movement and has suggested that one motive for support was to reorganize agriculture in nonpolitical ways that would eliminate radical elements remaining from earlier days of the Populist threat to trust and monopolies. These interests, aligned as the National Soil Fertility League worked to secure passage of the Smith-Lever Extension Act, an act that tightened the influence of business and its allies in the USDA over the university agricultural community.

Growing accountability to the USDA only increased the marginal status of mountain farming. J. Russell Smith, one of the earliest advocates of a diversified and regionally specific approach to university and national farm planning, wrote in 1916 that mountain farms "are the victims of an economic tragedy—the attempt to practice level-land agriculture on the unmitigated mountainside"(1916:330). He described the mountain climate as favorable— a "Garden of Eden" with heavy rainfall, an elevation fostering early springs and late autumn frosts, and deep and fertile soils. Smith had traveled around the world studying mountain agriculture and concluded that in Corsica, which had "less favored slopes," a "genuine mountain agriculture has been developed"(1916:331). He went on to claim that "the task of the schools of the great organizations that we have built up for the dissemination of agricultural knowledge" was to create "good little permanent terraced fields" for farming in Appalachia. Smith concluded with an attack on emerging farm policies: "We have a Federal Department of Agriculture, many State departments, State colleges, State experiment stations, substations, and a host of peripatetic demonstrators. Can they not among them develop and teach a mountain agriculture that will make the mountaineer prosperous and leave him his mountain?" (1916:333).

Marshall Vaughn also criticized the USDA for single-mindedly applying a science of agriculture unsuited to the mountains. In 1925 he wrote that a goal for mountain farmers was "educating the National Department of Agriculture in the particular needs of the mountains along agricultural lines" (1925c:20). He went on to specifically attack an overemphasis on flatland Bluegrass farming: "Crops and methods of cultivation that fit the soil and the irregular surface of the land should be made use of. Fruits and garden products, sheep, milk-cows, and poultry are better suited to our conditions than grain, tobacco or race horses" (1925c:20).

It is important to consider that, even if university specialists in Lexington had focused on mountain farming during these early years, an entirely different conception of farming would have been required for them to understand and support mountain agriculture. Mary Verhoeff (1917:138) has pointed out that mountain farming and central Kentucky farming at the time of the Civil War were based on two quite different economies. Central Kentucky farmers operated large plantations based on slave labor and competed in a commercial agricultural market with farmers in the rest of the nation and in the world. In eastern Kentucky, although the same crops were raised, farming had not been for commercial purposes. In her study of Kentucky River navigation in the early 1900s, Verhoeff surveyed the wide variety of products raised in eastern Kentucky and reported:

> None of these staples have been produced in quantities more than sufficient for the home demand. The only important shipments from the farms have consisted of livestock. . . . None of the agricultural products are shipped by individual farmers. Instead, they are taken to the nearest store and when a sufficient amount has accumulated they are sent out in wholesale lots and now by rail. Formerly they made up a considerable part of the river cargoes (1917:140-142).

Eastern Kentucky agriculture was not an industry, then, at the time when the university programs were developing. Yet, increasingly, the new science of agriculture depended on industrial assessments of its own success. What Rosenberg called "the worship of productivity as the essence of and index to progress" (1976:141) characterized agricultural research from the beginning. Those farmers who backed the land-grant colleges wanted ways to increase their productivity and gain a competitive edge over other farmers. Increasing farm productivity was the primary farm problem agricultural scientists tackled in their new experiment stations. Eastern Kentucky farms were among the least productive in the state. From the beginning, then, university farm specialists considered mountain farms economically inferior. Consider the conclusions drawn by Dr. Willard Rouse Jillson, a state geologist, when he compared eastern and Bluegrass farms: "It is not anticipated that this area can ever be made even by the practice of the most modern and diligent methods, to compete successfully with the Bluegrass area adjoining"(1928:12). A similarly fatalistic attitude characterized Tennessee's extension work in its eastern mountainous farm districts. McAmis of the University of Tennessee's Extension Division felt that the same principles of productivity should hold in mountain areas as elsewhere and concluded: "Mountain lands can never compete with good lands; but since the people will not move out of the mountains to better farming sections, the only thing to do is to help them use what they have to the best advantage" (1926:24).

This focus on economic competition and high productivity among agricultural scientists tended to overshadow another form of productivity—the ability to produce those goods needed to support a self-sufficient farm population. The yardsticks of university farm specialists appeared to measure such farms as Henry Combs's farm in Breathitt County in the mid-1800s as "unproductive" and "marginal":

> He [Henry Combs] became one of the most prosperous and farsighted farmers in this region, taking up a large section of virgin land on Troublesome Creek, comprising about 2,000 acres with about 25,000,000 feet of timber on it. He cleared much of this land and raised fine crops. By 1870 Henry Combs had planted on his farm 1,000 apple trees and ten acres of peaches, together with two acres of nursery stock for grafting purposes. He grew his cotton, ginned it, and the women on the place spun and wove it into clothing for the family. He made his own brick from clay on his farm, and he claimed to have made the first fireplace in this section of Breathitt County. He tanned leather and made shoes for all his family, shaped on lasts made by himself. . . . Henry Combs also erected a schoolhouse on his place and hired a teacher (Writers Program, 1941:45-46).

Corn production offers another example of the consequences of measuring productivity in terms of high yields. Verhoeff (1917:138) reported that corn was the principal product in eastern Kentucky in the early 1900s, occupying at least three-fourths of the tilled land. However, in a discussion among university farm specialists in 1926, the conclusion was reached that "in the raising of corn the mountains cannot compete economically with corn raising in the corn belt" (McAmis, 1926:24). The greatest portion of corn raised by mountain farmers was for household use, followed by feed for livestock and the production of corn liquor (Verhoeff, 1917:138). Since this crop was not raised for commercial sale, university specialists were unconcerned about corn production and concluded that the home vegetable garden was the most important aspect of mountain farming (McAmis, 1926:24).

Still, some early extension work in eastern Kentucky recognized that mountain farmers could find ready markets locally because of industrialization. While competition in national markets seemed doomed, a mountain agriculture might flourish by supporting the industrial boom. An "Eastern Kentucky Poultry Improvement Project" started in 1924 was planned to improve flock quality, build and remodel poultry houses, keep records of flocks, teach programs of balanced feeding, and set up Junior Agricultural Clubs for boys and girls. Estill County Agent K.J. Bowles said that "the poultry industry is of greater importance for eastern Kentucky than any other single farm enterprise" (Price, 1927:12). Pike County Agent M. Willis Abner reported that the project encouraged banks to loan money to farmers since "poultry in Pike County brings the farmers more ready cash than any other single enterprise on the farm" (Price, 1927:14). The KAES felt that the project would increase nonindustrial income for families in coal areas since "some of the best markets in the country are in the mountains . . . mining camps and towns or small industrial centers" (Price, 1927:13). Similarly, the associate editor of the *Southern Agriculturist* argued, "Mountain markets are daily growing better. There is a demand for all kinds of things to eat at the mines, lumber camps, and saw mills, vacation resorts, summer camps and hotels" (Neel, 1927:6).

This interest in developing local markets and encouraging greater economic independence for mountain farmers was slight when compared with work going on at about the same time in a nearby mountain state. In North Carolina, where coal was not an issue, extension agents helped organize the "Farmers Federation," an elaborate distribution and marketing program. Realizing that a diversified approach was essential to mountain farming, the Federation organized a chain of warehouses, a milk sales organization, a trucking fleet, hatcheries, canning plants, orchard marketing, a sweet potato curing house, and forest products industries (Green, 1928; McClure, 1931).

Even before the Farmers Federation, North Carolina agricultural specialists were working on ways to make mountain farmers economically independent in a changing agricultural national economy. Between 1913 and 1916, 26 cooperative cheese factories were organized (Doane and Reed, 1917). These efforts were closely associated with the dairy industry and were probably designed to dampen farmers' protests. Still, they reflect a conscious and substantial commitment to mountain farming needs in North Carolina.

University of Kentucky specialists were clearly aware of the North Carolina programs. The "Conference Numbers" of *Mountain Life & Work* magazine during the last half of the 1920s and the 1930s report on agricultural discussions at gatherings that included extension agents and farm specialists from all the central Appalachian states. But because eastern Kentucky was so clearly defined as an industrial area, under development by coal and timber capitalists, ideas for mountain farming there remained crippled. Even the editors of the *Jackson Times,* who so enthusiastically supported industrialization, felt that the university was slighting mountain farming:

> Much has been written and said about the wonderful timber and mineral resources of Breathitt County, but strange to say, there has been little attention given the matter of its agricultural possibilities. . . . The proper encouragement is not given to farmers. Most of the farm stuff at present comes from outside with the added high costs of transportation. Breathitt should produce at least all it consumes. . . . There is no good reason why small garden and field crops fail to pay. As a fruit region, we predict the day is not far distant when such fruits as apples, pears, peaches, plums and others will form the mainstay of its farmers. Nor have we taken into consideration the peculiar advantages possessed for cattle raising and dairying. . . . With good mountain roads, such as Breathitt will sometime have and perhaps additional railroad facilities, agriculture will be considered one of the county's most valuable assets (*Jackson Times,* June 16, 1911).

Unfortunately, the Breathitt County newspaper was a bit more optimistic about mountain farming than many agricultural scientists, who made no concerted effort to tackle marketing and distribution of products of mountain farming.

The role of home economics extension in Kentucky and the nation was clearly supportive of industrialization. The coming of coal and timber development drastically changed daily life for mountain women. The State Commissioner of Agriculture in his 1887 Annual Report called attention to the important role women played in eastern Kentucky before industrialization. Referring to Perry County, he said,

> There are but few manufactures in this county, except what are conducted by the economical, intelligent, industrious women who keep up the good old fashion which they learned from their mothers and grandmothers, of using the flax-wheel and distaff, the cotton and wool-cards, the wheel, the warping-bars and the hand-loom. Every girl is taught how to make the cross and beam the warp, to use the treadle and ply the shuttle to excel in making linsey, jeans, cottonades, linen, counterpanes, and huckaback towels and bed-spreads. After supplying the family with these articles the good ladies traffic the surplus as well as large amounts of beeswax, butter, eggs, poultry, feathers, home-knit socks, dried fruit, maple sugar, tree molasses, and ginseng to the county merchants in exchange for "store goods," "spun truck," fancy dresses, ribbons, jaunty hats, hoop-skirts, groceries and whatever other articles may be essential for their comfort and Sunday apparel (cited in Verhoeff, 1917:142-43).

Three to four decades later, however, these women had been moved off their farm homesteads and found themselves crowded into coal-camp houses or lumber-camp shacks. Industrialization had stripped women of many of the resources with which they so skillfully supplied their family's needs. In the wake of these changes, home economics extension workers came into the mountains. On Agricultural Trains, for instance, two whole cars were devoted to exhibits and lectures tailored to women. Smith called these extension workers "pioneer home agents." They taught women in coal and lumber camps "nutrition . . . home management, clothing construction and care" and helped set up girls' food clubs and women's clubs (1965:38-39). The homemakers' clubs which these extension agents helped organize were actually designed to "modernize" women's work. Success to these agents meant encouraging uprooted mountain women to adopt a role congruent with the new national stance of woman-as-consumer.*

The E.O. Robinson Substation.

It is significant that eastern Kentucky extension began in earnest only when timber industrialists donated land to the state. E.O. Robinson of Fort Thomas, Kentucky, and F.W. Mowbray of Cincinnati had operated one of the largest lumber mills ever located in eastern Kentucky. From 1908 to 1922 the Mowbray and Robinson Co. had feverishly cut down the great timber stands on 15,000 acres in Breathitt, Perry, and Knott counties. Nevyle Shackelford, "information specialist" for the College of Agriculture, described the timbering:

> Great trees that had been growing since before Columbus discovered America were hastily cut down, sawed into logs, loaded unceremoniously on flat cars—and later, without respect to their age, beauty or hold on history, reduced to lumber that was used for the construction of everything from mansions to pig pens and from sailing ships to packing crates. . . . By [1922] all the giant oak, poplar, beech and other patriarch trees had been lopped from the coves and the steeps of the mountainsides. It was then that Mowbray and Robinson pulled up stakes and departed (1979:10-11).

When Robinson set aside a fund of $1 million and 15,000 acres of cut-over timber for the state to use, he told a *New York Times* reporter: "The mountains have given much to me; it is little I am giving back to them" (*New York Times,* 1923).

In 1924 the General Assembly accepted the land and created an experiment station for the mountains, naming it the "Robinson Substation." For two years the state "vacuumed" the debris left from the timbering operation—burning the huge sawmills and wood scraps and razing the shacks that had housed mill workers. The Robinson gift insured university attention to mountain farming for the first time. Dean Thomas Cooper commented, "We have hopes for the economic development of this vast region as a result of the creation of the Robinson fund on a scale that never before had been

*Stuart Ewen (1976) has noted that the development of home economics nationally is intimately tied to industrialization and the expansion of capitalism.

practicable"(*New York Times:*1923). An outright gift of $1 million and 15,000 acres, no matter how cut-over, could hardly be ignored. In his 1924 Annual Report Dean Cooper said, "Mountain agriculture, the prosecution of work in forestry and the varied demonstrations of the possibilities of eastern Kentucky can only be carried on in this region" (Smith, 1965:26). At last the university was tackling mountain farming head on. Still, the date of the concern and the tone of the work was determined by the economic order which had stripped the lifeblood from the farm population. After cleaning up, the university worked most intensively on reforestation. Trees in eastern Kentucky and the whole of the central Appalachian highlands had been cut over by timber barons; reforestation was a crucial problem.

When the Robinson fund was announced, Breathitt Countian R.T. Gunn wrote with enthusiasm about possibilities for university work in the mountains: "Here is an opportunity for the University of Kentucky to do two jobs, teach us to raise tie timber for the Railways, props for the mines, and that is going to be one of the great problems up here soon, and other valuable timber also. Teach us to make the best use of the wonderfully located bottom lands along the Kentucky River" (1923). Staff at the Robinson Substation did work to improve farming methods for what remained of bottom-land farms. However, they had little interest in much of the former farm population, suggesting that mountain lands should be bought by the state or federal government for reforestation work and that social workers should encourage people to "seek better farming sections" (McAmis, 1926:24-25). As Professor Anderson of the University said, "The great objection to this plan is, however, that the people love the mountains and in spite of handicaps wish to remain in them" (McAmis, 1926:25).

In 1935, the first comprehensive survey of central Appalachia was released. Prepared in cooperation with the Office of Education, U.S. Department of Interior, and the Agricultural Experiment Stations of Tennessee, Virginia, West Virginia, and Kentucky, the report recommended that much of the mountains should be turned over to state and federal ownership, that intentional rural communities should be set up near forest industries and mining areas, that some local manufacturing should be encouraged, and that the people of the region should be encouraged to emigrate (USDA, 1935:6). To these professional, well-intentioned people, a primary solution for mountain farmers involved "regrouping" the population along two major lines. First, part-time farming would be combined with primary employment in the forests and forest industries. Second, part-time farming would be combined with primary employment in coal mining (USDA, 1935:4-5). The solutions that the USDA-backed team envisioned for a population whose agricultural support base had been destroyed by coal mining and timbering were either to move the people closer to coal and timber jobs or to move the people out. These recommendations were proposed as "a rational utilization of our agricultural resources" (USDA, 1935:2). The underlying philosophy among agricultural scientists regarding mountain farming was that the mountains were "incapable of supporting a farm population" (USDA, 1935:3). Therefore, they hoped to encourage farmers to leave and, when that failed, to offer techniques designed for use elsewhere, seed varieties, fertilizers, and other methods that might modernize mountain farming. They did not challenge absentee ownership of the land, corporate land abuse, the lack of credit for small mountain farmers, the absence of a technology appropriate for the region, and the dependence of the mountain economy on coal and timber expropriation by outside industrialists. All this is not to say that the work at Robinson Substation did not help mountain farmers. Much of it did. New strains were introduced into hen flocks, dairy and beef cattle herds, and hogs. Years of industrial activity had interrupted the genetic upgrading that had occurred from both the livestock drives through the mountains to southern and eastern markets in the early and mid-1880s and from introduction of hybrid stock by more prosperous farmers (Verhoeff, 1917:140, 146). In its first year, the substation set up a trade of one bushel of "improved seed corn" to every farmer who brought in a bushel of "ordinary corn"; a purebred rooster was given "for any old mongrel rooster" brought in; an "improved boar for any old razorback" (Shackelford, 1979:13). Demonstrations of methods of farming, gardening and fruit raising, and care of woodlands were also begun. By 1926 the substation had dairy, hog, and poultry breeding programs underway, based on research carried out earlier by the KAES and with an aim of producing new research suited to mountain needs (Schackelford, 1979:14-16).

Just as experiment-station workers in Lexington were under pressure to justify their programs by getting information out to farmers, the Robinson Substation workers needed to disseminate their findings and encourage farmers to adopt techniques they recommended. In 1926 the substation initiated a "Robinson Harvest Festival," which drew people from about a 40-mile radius (Shackelford, 1979:17). This festival became the "biggest annual event in Breathitt County" (Writers Program, 1941:144). The event, which continued until World War II, became a sort of reaffirmation of farm life. People brought exhibits of their crops and handicrafts for display, and frequently the old barter system of agricultural trade was renewed for the duration of the festival.

However, most of the new farming methods advocated by the KAES were imported wholesale to eastern Kentucky. As McConnell has commented about the spread of new technology among agricultural workers, "the new machines, plants, fertilizers, and all the new developments were looked on as undiluted goods" (1969:14). C.J. Galpin of the Bureau of Agricultural Economics of the USDA, for example, wrote that much of the toil of farming in the mountains could be removed through "the machine in the farm work of the man, the machine in the household work of the woman . . . the slaves of the farmer and of the farm woman" (1927:3). Yet, as L.R. Neel pointed out, "the tractor and large tillage tools are to a great extent precluded" in the mountains (1927:4). Even for those farmers who still owned bottom-land farms, credit was so scarce for mountain farmers that buying the suggested machinery was next to impossible. In a 1933 study of factors affecting farm credit in Kentucky, the KAES reported that all farmers in the study had received credit from outside sources like insurance companies and federal and joint-stock land banks *except* those farmers in the study from mountain areas. For mountain farms, only local banks and individuals held farm mortgages (Bradley, 1933:153-88).

Other recommendations made by Substation workers were financially difficult for farmers. In 1890, for instance, there were no expenditures for fertilizers in Clay, Knott, Lee, Leslie, Owsley, and Perry counties (Verhoeff, 1917:145). Most farmers relied on manure. Extension workers, in an effort to introduce scientific techniques, distributed bulletins about the advantages of various types of fertilizers. In the years of unemployment following the closing of the sawmills, few people had money for such improvements (Duerr et al., 1946:16-26), even if they still lived on their own land.

The early extension workers in the mountains had considerable prestige in the new social hierarchy that followed industrialization. Vaughn described extension agents as "the real leaders in most of the Eastern Kentucky counties" (1925d:14). They encouraged farmers to do such things as make home repairs, build new stock barns, plan home lighting plants, build community creameries, rejuvenate worn soil, build wire fences, plant trees, improve chicken flocks, and increase the market value of vegetables. James M. Feltner was an eastern Kentucky field agent for Junior Clubs in the Extension Service in 1918 who became one of the better-remembered early extension workers. The great-grandson of one of Leslie County's earliest settlers, Feltner traveled by horse and by foot through many mountain counties. Vaughn described his work as follows.

> Mr. Feltner's principal work consists of organizing community farm practices and home life. This includes poultry, sheep, pigs, calves, corn, potatoes, canning, cooking, sewing and other projects that are suitable for permanent development in a given locality. He gives the youth scientific methods of doing the work, and backs it up with a beautiful spiritual philosophy that puts a soul into the enterprise (1925b:27).

County agents and extension workers, then, saw their roles as changing traditional ways of farming in the mountains and bringing in techniques of scientific farming which they learned at the university. Defining mountain agriculture as subsistence by nature, rather than as subsistence as a result of industrialization, the agricultural scientists and extension workers focused on helping mountain families adapt to an economy over which they had no control and which periodically in its boom-bust fluctuations left them near destitution.

CONCLUSIONS

Eastern Kentucky provides a case study of the role of agricultural science in the demise of a farm economy. This study has important implications for both the current critiques and the current practices of university agricultural research and extension work.

Critics of the close ties of the land-grant colleges to agribusiness frequently tend to be ahistorical. Wendell Berry, for instance, offers a severe indictment of the land-grant colleges, charging that their programs are so subjugated to the interests of agribusiness that they have become a key force in the "unsettling" of American small farmers. Berry (1977) cites the 1955 Amendment to the Smith-Lever Act (Section 347a) as a turning point, suggesting that prior to the amendment there was adequate legislative backing for agricultural scientists to devise a small-farm technology. He implies that somewhere along the way sentiment shifted from building colleges in support of the nation's rural population to building a complex that supported corporate agriculture. The case of eastern Kentucky indicates that the roots of the agribusiness/university relationship can be traced far back before 1955. To understand the nature of that relationship, specific historical and comparative research is needed on the creation of the land-grant programs as they emerged in each location across the country.

A grasp of the economic and social realities in which each college forged its programs would extend the critique of the overall complex. Busch (1980) has argued that the contemporary structure of agricultural science research is the result of a "negotiation process." Interest groups as well as scientific traditions are involved in the formulation of research strategies by agricultural scientists. That process occurs within the structural relationships established between universities and state and federal legislatures. If we examine the process of "negotiating" research and extension strategies within structural and temporal contexts, the critique of agricultural science becomes more complex. It reaches beyond an attack on big-business manipulation that dominates the Hightower and DeMarco (1973) indictment. That critique, important though it is, needs to be extended so that we understand how people working in agricultural colleges, with intentions of helping farmers and riding on the idealism of a new federal program of people's universities, came to build a university network that has made possible a rapid decline in farm population. Eastern Kentucky offers an example of the way in which agricultural scientists came to define one group of farmers as marginal.

The fact that mountain farmers were trying to survive in an area undergoing intense industrialization pushes the critique to another level. In those counties where resource extraction has dominated local economies, the key piece of the puzzle of the demise of farmers is the incorporation of those areas into an expanding capitalist economy. To many people, the demise of mountain farming in eastern Kentucky appears to be an inevitable aspect of progress and development. But comparative research on farm programs in other mountain areas in which coal did not figure so prominently, as in North Carolina, exposes the fallaciousness of such notions. The eastern Kentucky experience reveals that the research conducted by agricultural scientists tended to support the class interests of capitalists who, at that point in time, were searching for new sources of raw materials and new investment potential.

From the late 1800s on, the extraction of resources in eastern Kentucky has determined the shape of the local economy. The situation is not likely to change. Even prior to the energy crisis of the 1970s and its boost to coal production, the central Appalachian coalfields produced some 70 percent of the nation's coal. Furthermore, coal reserves are estimated to be able to supply the national energy demand for two hundred years or more (Gaventa, 1975:44). As demand for coal, timber, and other mineral resources has fluctuated, eastern Kentuckians have become "economic refugees," leaving the area in search of jobs in midwestern industrial cities and returning when those jobs dried up or with the emergence of a coal boom (Tudiver, 1977; Council of the Southern Mountains, 1976; Maloney, 1972a, 1972b). Still, researchers and extension workers have failed to recognize this basic characteristic of eastern Kentucky life.

The focus of extension in eastern Kentucky has been on changing the behavior of individual farmers. According to Shackelford, "Because of information made available by the Cooperative Extension Service and by personal observances, farmers adopted new strains of crops, changed their methods of land use and farming practices . . . . It can be truthfully said that because of Robinson Substation, life in the region has become infinitely better, neater, more comfortable, more progressive, and more profitable" (1979:28).

The university's annual field trials indicate how extension often works today in the mountains. Yearly, on-the-farm demonstration work tends to occur on the largest, most highly capitalized farms. In 1978, for example, an Owsley County farm that is reputed by old-timers to be the best farm "anywhere around" was chosen. The owners, a small family corporation, are in a better financial position than most small farmers in the demonstration area. They are more likely to obtain the credit needed to make improvements in field restoration, water distribution, construction of farm buildings, purchase of machinery, etc. They are, therefore, most likely to increase productivity. As Arndt and Ruttan point out, productivity has become "a useful device for monitoring research program performance" (1977:8). As a result, extension workers tend to select those mountain farms most likely to adopt new techniques and increase productivity.

This approach is justified by much of the literature on rural development. Since the early 1960s, research on the "adoption and diffusion of innovations" has dominated notions of development. Rogers and Shoemaker (1971), for instance, suggest that a sort of "natural" selection occurs when new agricultural innovations are introduced. Modern, cosmopolite, science-oriented, less dogmatic, businesslike, empathetic, socially mobile individuals adopt new techniques. Less progressive-minded farmers may eventually copy them, in a sort of trickle-down effect. This, then, justifies an almost exclusive focus on farmers in the best position to try out recommendations. Those smaller farmers who fail are seen as unfit due to personal inadequacies. The case of the demise of most of the farm population in eastern Kentucky exposes the limited nature of this individual rationale upon which—it is claimed—farmers succeed or fail. Still, much of extension work is based on the diffusion approach.

Attempting to get coal or timber corporations to change *their* behavior and "methods of land use" has never been an issue. Instead, the university came into eastern Kentucky to salvage what was left in the mountains when resource extraction hit a lull. As Landy has shown, the university responded to Governor Breathitt's efforts to control strip mining "only when its own forest research facility was being threatened" (1976:332).

Those who advocate an "institution-building" approach to development also fail to see that the failure of many small farmers—in both the United States and abroad—is related to developments in an international economic system. No underdeveloped area can support "modern agriculture" without what I.L. Baldwin calls an "adequate infrastructure" (1972:26). This means that there is a need for some sort of technical assistance facility to create solutions relevant to local problems. It should be staffed by "native" agriculturalists who would learn their skills through participant training programs. Working through local elites trained in "modern" farm methods to help change the behavior of "traditional" farmers misses the basic point made above: the local elite in eastern Kentucky tended overwhelmingly to support the industrialization that destroyed the agricultural land base. The rationale underlying the institution-building approach resembles the 1961 "Eastern Kentucky Resource Development Project," initiated by the university with the help of a sizable financial grant from the Kellogg Foundation. J. Allen Smith wrote: "The area, along with other areas of the Appalachian region, had become a 'disadvantaged area,' largely *by-passed by the economic and institutional development* that had occurred in other parts of the nation" (1965:43; emphasis added). The problem mountain farmers faced was not that they had been "by-passed" by industrialization. It was the particular way in which their region had been drawn into the world capitalist economy.

Saint and Coward (1977) have suggested that an emerging interest among behavioral scientists is affecting current agricultural research activities. New attention is being focused on technology as a social product, the research process as a subject of study itself, questions of alternative technologies, social consequences of particular technologies for various situations, and the importance of indigenous knowledge. As these new ideas gain more recognition, some changes in official agricultural policies may occur. But, as the eastern Kentucky case shows, the grounding of agricultural research and extension has in the past been fundamentally tied to a business outlook and world view that helps maintain the dominance of the world's energy and food corporations in the international class structure.

REFERENCES

Alinsky, Saul. 1970. *The Unauthorized Biography of John L. Lewis.* New York: Vintage Books.

American Association of Land-Grant Colleges and State Universities. 1962. *Land-Grant Fact Book: Centennial Edition.* Washington, D.C.: Centennial Office of the American Association of Land-Grant Colleges and State Universities.

Appalachian Regional Commission. 1968. *Appalachia* 1 (August). Washington, D.C.: Appalachian Regional Commission.

_____. 1972. *Appalachia—An Economic Report: Trends in Employment, Income, and Population.* Washington, D.C.: Appalachian Regional Commission.

_____. 1975. *Annual Report.* Washington, D.C.: Appalachian Regional Commission.

Arndt, Thomas M., and Vernon W. Ruttan. 1977. Valuing the Productivity of Agricultural Research: Problems and Issues. In *Resource Allocation in National and International Agricultural Research,* ed. Thomas M. Arndt, Dana G. Dalrymple, and Vernon W. Ruttan, pp. 3-25. Minneapolis: University of Minnesota Press.

Arnett, Douglas O. 1978. Eastern Kentucky: The Politics of Dependency and Underdevelopment. Ph.D. dissertation. Duke University.

Baldwin, I.L. 1972. Meeting the Needs for Professional Agriculturalists. In *Institution Building: A Model for Applied Social Change,* ed. D. Woods Thomas, Harry R. Potter, William L. Miller, and Adrian F. Aveni, pp. 25-39. Cambridge, Mass.: Schenkman Publishing Co.

Banks, Alan J. 1979a. *The Growth of a Working Class in East Kentucky, 1870-1930.* Unpublished manuscript. Hamilton, Ontario: McMaster University.

_____. 1979b. The Emergence of a Capitalistic Labor Market in East Kentucky, 1870-1915. Paper presented at the annual Appalachian Studies Conference, Jackson's Mill, West Virginia, March 16-18.

Berry, Wendell. 1977. *The Unsettling of America: Culture and Agriculture.* San Francisco: Sierra Club Books.

Bowman, John B. 1868. Report of the Agricultural and Mechanical College of Kentucky. Made to the Governor of the State of Kentucky by J.B. Bowman, Regent of Kentucky University, Dec. 28, 1868. Frankfort, Ky.: Commonwealth of Kentucky.

_____. 1872. Report Concerning the Agricultural and Mechanical College of Kentucky by the Regent of Kentucky University to Governor P.H. Leslie, Feb. 7, 1872. Frankfort, Ky.: Commonwealth of Kentucky.

Bradley, C.J. 1933. The Use of Credit in Selected Farms. University of Kentucky Agricultural Experiment Station Bulletin 343 (June): 153-188. Lexington, Ky.: Kentucky Agricultural Experiment Station.

Bralliar, Floyd. 1927. Fruit on the Mountain Farm. *Mountain Life and Work* 3 (April): 17-18, 25.

Bryant, Tommy R. 1911. Forestry for the School and the Farm, Newspaper Bulletin no. 15, (October 23). Lexington, Ky.: Kentucky Agricultural Experiment Station.

Busch, Lawrence. 1980. Structure and Negotiation in the Agricultural Sciences. *Rural Sociology* 45:26-48.

Campbell, John C. 1969. *The Southern Highlander and His Homeland.* Lexington, Ky.: University Press of Kentucky. Reprint of 1921 edition by the Russell Sage Foundation.

Caudill, Harry M. 1962. *Night Comes to the Cumberlands: A Biography of a Depressed Area.* Boston: Little, Brown & Company.

Childers, Joey. 1977. Absentee Ownership of Harlan County. Unpublished report to the Appalachian Center. Lexington, Ky.: University of Kentucky.

Council of the Southern Mountains. 1976. Special Issue: Urban Appalachians. *Mountain Life and Work* 52 (August).

Craig, Ronald B. 1932. Forestry in the Economic Life of Knott County, Kentucky. University of Kentucky Agricultural Experiment Station Bulletin 326 (February):1-39. Lexington, Ky.: Kentucky Agricultural Experiment Station.

Doane, C.F., and A.J. Reed. 1917. Cheesemaking Brings Prosperity to Farmers of Southern Mountains. *USDA Yearbook, 1917:*147-152. Washington, D.C.: United States Department of Agriculture.

Dreiser, Theodore. 1970.Introduction. In *Harlan Miners Speak: Report on Terrorism in the Kentucky Coal Fields.* Ed. Leonard W. Levy, pp. 3-16. New York: Da Capo Press. Reproduction of original 1932 edition by Harcourt, Brace & Co.

Duerr, William, John B. Roberts, and R.O. Gustafson. 1946. Timber-products marketing in eastern Kentucky. University of Kentucky Agricultural Experiment Station Bulletin 488 (June):1-95. Lexington, Ky.: Kentucky Agricultural Experiment Station.

Eller, Ronald D. 1976. Industrialization and Social Change in Appalachia, 1880-1930: A Look at the Static Image. Paper presented at the Southern Historical Association, Atlanta, Ga.

———. 1977. The Coal Barons of the Appalachian South, 1880-1930. *Appalachian Journal* 4 (Spring-Summer):195-207.

Ewen, Stuart. 1976. *Captains of Consciousness: Advertising and the Social Roots of the Consumer Culture.* New York: McGraw-Hill.

Gabbard, Mrs. Myrtle Ward. 1979. Personal interview in Berea, Ky, April 15.

Galpin, C.J. 1927. Saying a Thing or Two for the Mountains. *Mountain Life and Work* 3 (April):1-3.

Gaventa, John. 1973. In Appalachia, Property Is Theft. *Southern Exposure* 1 (Summer-Fall):29-41.

———. 1975. Power and Powerlessness: Quiescence and Rebellion in an Appalachian Valley. Doctor of Philosophy Thesis. Nuffield College, Oxford University.

Green, Zeb. 1928. The Farmers Federation. *Mountain Life and Work* 3 (January): 30-31.

Gunn, R.T. 1923. The Robinson Mountain fund. *Jackson Times* (April 27):1.

Gunning, Sara Ogan. 1979. Personal interview in Lexington, Ky., April 16.

Hadwiger, Don F. 1975. The Green Revolution: Some New Perspectives. *Change* 7 (November): 36-41,62.

Hightower, Jim, and Susan DeMarco. 1973. *Hard Tomatoes, Hard Times: A Report of the Agribusiness Accountability Project on the Failure of America's Land Grant College Complex.* Cambridge, Mass. Schenkman Publishing Co.

Hopkins, James Franklin. 1938. A History of the Hemp Industry in Kentucky. M.A. Thesis. Lexington, Ky.: University of Kentucky.

Jackson Times. 1911. *Jackson Times.* Various issues.

———. 1923. Buckhorn Nursery Company. *Jackson Times* (June 22):7.

Jillson, Dr. Willard Rouse. 1928. Geology of Eastern Kentucky Soils. *Mountain Life and Work* 4 (October):11-13,17.

Kentucky Agricultural Experiment Station. 1911-15. Kentucky Newspaper Bulletins. Numbers 2-119. Lexington, Ky.: Kentucky Agricultural Experiment Station.

Kirby, Richard. 1969. Kentucky Coal: Owners, Taxes, Profits: A Study in Taxation without Representation. *Appalachian Lookout* 1 (6):19-27.

Landy, Marc Karnis. 1976. *The Politics of Environmental Reform.* Washington, D.C.: Resources for the Future.

Levy, Melvin P. 1970. Class War in Kentucky. In *Harlan Miners Speak: Report on Terrorism in the Kentucky Coal Fields,* ed. Leonard W. Levy, pp. 20-49. New York: Da Capo Press. Reproduction of 1932 edition by Harcourt, Brace & Co.

McAmis, Mr. 1926. Round Table on Agriculture. *Mountain Life and Work* 2 (July):24-25.

McClure, James G.K., Jr. 1931. Ten Years of the Farmers Federation. *Mountain Life and Work* 7 (April):23-25.

McConnell, Grant. 1969. *The Decline of Agrarian Democracy.* New York: Atheneum. First published in 1953 by University of California Press.

McConnell, John P. 1929. The Retardation of the Appalachian region. *Mountain Life and Work* 5 (April):21-22.

Maggard, Mary Ward. 1979. Personal interview in Berea, Ky., April 15.

Maggard, Sally Ward. 1977. Local Newspapers in Eastern Kentucky: A Profile of an 8 County Area. Unpublished manuscript. Lexington, Ky.: University of Kentucky.

Maloney, Michael E. 1972a. Migration: A Problem of Community. *People's Appalachia* 2 (July): 4-5.

———. 1972b. Appalachian Settlements: Cincinnati and Southwestern Ohio. *People's Appalachia* 2 (July):29-30.

Middlesboro Daily News. 1965. Anniversary Special Edition. *Middlesboro Daily News* 55 (August 19).

Millstone, James. 1972. East Kentucky Coal Makes Profits for Owners, Not Region. In *Appalachia in the Sixties,* ed. David Walls and John B. Stephenson, pp. 69-75. Lexington, Ky.: University of Kentucky Press.

Murray, Thomas J. 1965. The Investment Nobody Knows About. *Dun's Review and Modern Industry* 85 (April):40-43.

Neel, L.R. 1927. Agriculture in the Southern Mountains. *Mountain Life and Work* 3 (April):4-6,10.

New York Times. 1923. Gives Millions to Aid Southern Mountaineers. *New York Times* (September 9): section 7, page 8.

O'Donnell, Erin. 1979. Plan Seeks Corporate Land for House Sites. *Herald-Voice* (June 1):1.

Pilcher, Louis. 1913. *The Story of Hazard, Kentucky, the Pearl of the Mountains.* Hazard, Ky.: Hazard Herald.

Pope, Liston. 1942. *Millhands and Preachers.* New Haven, Conn.: Yale University Press.

Price, Charles S. 1927. Poultry in the Hill Country. *Mountain Life and Work* 3 (April):11-14.

Raine, James Watt. 1924. *The Land of Saddle-bags: A Study of the Mountain People of Appalachia.* New York: Council of Women for Home Missions and Missionary Education Movement of the U.S. and Canada.

Ridgeway, James. 1973. *The Last Play.* New York: E.P. Dutton.

Roberts, George. 1944. *A Brief History of the Development of Work in Agronomy in the College of Agriculture and Home Economics.* Lexington, Ky.: University of Kentucky Press.

Rochester, Anna. 1970. Who Owns the Mines? In *Harlan Miners Speak: Report on Terrorism in the Kentucky Coal Fields,* ed. Leonard W. Levy, pp. 50-58. New York: De Capo Press. Reproduction of 1932 edition by Harcourt, Brace & Co.

Rogers, Everett M., and F. Floyd Shoemaker. 1971. *Communication of Innovations: A Cross-Cultural Approach.* New York: Free Press.

Rosenberg, Charles. 1976. *No Other Gods: On Science and American Social Thought.* Baltimore, Md.: Johns Hopkins University Press.

Saint, William S., and E. Walter Coward, Jr. 1977. Agriculture and Behavioral Science: Emerging Orientations. *Science* 197 (August 19):733-37.

Shackelford, Nevyle. 1979. *Robinson Substation: A Short History.* Lexington, Ky.: University of Kentucky College of Agriculture Cooperative Extension Service.

Smith, J. Allan. 1965. *The College of Agriculture and Home Economics, University of Kentucky.* Lexington, Ky.: University of Kentucky Centennial Committee.

Smith, J. Russell. 1916. Farming in Appalachia. *American Review of Reviews* 53 (March):329-36.

Tudiver, Sari. 1977. Country Roads Take Me Home: The Political Economy of Wage-Labour Migration in an Eastern Kentucky Mountain Community. Unpublished manuscript. Available through author at Department of Anthropology, University of Manitoba, Winnipeg, Canada.

Turner, Wallace B. 1954. Kentucky in a Decade of Change. Ph.D. Dissertation. Lexington, Ky.: University of Kentucky.

United States Department of Agriculture. 1902. Message from the President of the United States Transmitting a Report of the Secretary of Agriculture in Relation to the Forests, Rivers, and Mountains of the Southern Appalachian Region. 57th Congress, First Session, Senate Document 84, Washington, D.C.: U.S. Government Printing Office.

———. 1935. Economic and Social Problems and Conditions of the Southern Appalachians. Miscellaneous Publication no. 205 (January):1-184. Washington, D.C.: United States Department of Agriculture.

Vaughn, Marshall E. 1925a. The Purpose of This Magazine. *Mountain Life and Work* 1 (April): 2-4.

———. 1925b. A Human Dynamo—Jim Feltner. *Mountain Life and Work* 1 (July):25-27.

———. 1925c. A Program for the Mountains. *Mountain Life and Work* 1 (April): 20-22.

———. 1925d. County Achievement Contest in Kentucky. *Mountain Life and Work* 1 (April):14-19.

Verhoeff, Mary. 1917. *The Kentucky River Navigation.* Filson Club Publications no. 28. Louisville, Ky.: John P. Morton and Company.

Wallerstein, Immanuel. 1974. *The Modern World-System: Capitalist Agriculture and the Origins of the European World-Economy in the Sixteenth Century.* New York: Academic Press.

———. 1976. From Feudalism to Capitalism: Transition or Transitions? *Social Forces* 55 (December):273-83.

Walls, David S. 1976. Central Appalachia: A Peripheral Region in an Advanced Capitalist Society. *Journal of Sociology and Social Welfare* 4 (November):232-47.

Weller, Jack. 1972. Appalachia's Mineral Colony. Vantage Point, no. 2. Reprinted in *Colonialism in Modern America: The Appalachian Case,* ed. Helen M. Lewis, Linda Johnson, and Don Askins. Boone, N.C.: Appalachian Consortium Press. 1978:47-55.

Worthington, William. 1978. Personal interview in Clintwood, Va., January 15.

Wright, Warren. 1970. The Big Steal. Paper prepared for the Council of the Southern Mountains. Reprinted in *Colonialism in Modern America: The Appalachian Case,* ed. Helen M. Lewis, Linda Johnson, and Don Askins. Boone, N.C.: Appalachian Consortium Press. 1978:161-75.

Writers' Program of the Work Projects Administration in the State of Kentucky. 1941. *In the Land of Breathitt.* Northport, N.Y.: Bacon, Percy and Daggett.

Part II

Science, Scientists, State: The Evolution of Agricultural Technology

3

Agricultural Research as
State Intervention

Christopher Dale

INTRODUCTION

This chapter examines agricultural research as a form of state intervention in capitalist society. To this end an attempt is first made to define the term "state" and to suggest that the basic role of the state in advanced capitalist nations is to serve dominant class interests. Brief attention to the state's function in American economic history offers some empirical support to the theoretical framework outlined.

Working from this general orientation, it is argued that agricultural research in the United States, initiated and largely funded by the state, has functioned so as to benefit prevailing economic interests involved in the agricultural marketplace. Support for this point is mustered from current literature on agricultural research, but focuses in particular on an historical analysis of the position of the Kentucky Agricultural Experiment Station vis-a-vis the Kentucky burley tobacco farmer over the years, 1885-1940. More specifically, the empirical analysis reveals that the experiment station buttressed agribusiness interests in tobacco products by virtually ignoring or reifying situations of market instability and intensely competitive production confronting the tobacco farmer. It concludes that the station's structural position amidst economic forces of the time fundamentally obligated it to support capitalist agricultural development.

THEORETICAL BACKGROUND

Broadly speaking, the term "state" refers to a whole configuration of institutions and organizations that formulate and/or enforce official public policy at the national, state, and local levels. It includes such elements as the executive, legislative and judicial branches of the government, the multitude of policy groups (regulatory commissions, advisory committees, research councils, etc.) that directly and indirectly channel political decisions, and the military and police apparatus. Based on this definition, most theorists would agree that the state has tended to assume an increasingly prominent role in the economy of capitalist nations, but disagree as to whether state intervention is essentially class-based or pluralistic in nature. That is, debate persists on whether a relatively cohesive ruling elite or conflicting, competing power blocs forge the contours of the political arena in capitalist countries.

As it is presented in the United States, the pluralist perspective holds that "many groups advance their claims upon government, each group receiving some of what it wants and no group receiving everything" (Best and Connolly, 1976:9). While it may be acknowledged that an upper class tends to provide the country or a community with most of its political leaders, it is also argued that the interests of this class are fragmented, and by no means coordinated. A majority of laypersons and "most social scientists, being of pluralist persuasion, believe that many different groups, including organized labor, farmers, consumers, and middle-class environmentalists, have a hand in political decisions" (Domhoff, 1978:7-8). Pluralism suggests that a matrix of political "checks and balances" insures that the state is essentially responsive to the general public, and that the "national interest" is served. From this perspective, the state's societal function "is envisioned as arbiter, stabilizer, and readjuster, trying as best it can to maintain stability while allowing for gradual (ameliorative) change" (Parenti, 1978:28).

The class-interest perspective emphatically denies the pluralist assumption that the state functions as a relatively "neutral" advocate and defender of the public interest. Rather, it sees the state as playing two crucial, interrelated roles: (1) maintaining social stability—i.e., a private profit system that preserves an inequitable distribution of power and wealth—and (2) stimulating the growth and proliferation of capitalist enterprise. These two general functions can be clarified by a brief reference to American political and economic history.

The class-interest perspective of power seems to offer a more accurate, convincing assessment of the state's position in capitalist society. Growing largely out of the political sociology of Karl Marx, this perspective points to the state as an increasingly prominent instrument of ruling-class domination over unpropertied classes. Domination is perpetuated because wealthier individuals (1) tend to occupy the key political offices in the nation or (2) have the financial muscle to influence the decision-making process in their favor (see Domhoff, 1978; Parenti, 1977, 1978). From the class-interest perspective, the many groups and organizations that dot the American political landscape project an illusion of conflicting interests among those who rule and widely divergent loyalties among the citizenry, even if that illusion is thinly veiled as in the case of "choosing" between a Republican and a Democratic candidate for office. Looking beyond this apparent multiplicity of political cleavages and factions, "the proof of the pudding in terms of power is the ability to maintain the class system that sustains ruling-class privileges and perogatives" (Domhoff, 1978:119). In this sense, it is argued, the ruling class has maintained a decisive advantage in perpetuating its interests via the state.

STATE INTERVENTION IN THE AMERICAN ECONOMY

It is commonly held that an American commitment to the regulation of capitalist market mechanisms with the not-so-invisible hand of the state first emerged in the late 1800s with legislation such as the Sherman Anti-Trust Act (1890) and began to take full form in the post-Depression era with the proliferation of New Deal policies under the Roosevelt administration. A major problem with this assumption is that laissez-faire capitalism per se never operated in the United States. "State intervention in every aspect of economic life is nothing new in the history of capitalism" (Miliband, 1969:9); in America it can easily be traced back to the colonial period. For example, Lichtman (1966:18) notes that "the Puritans brought with them [to America] the doctrine that the state had the right to regulate wage, price, service, and matters of similar concern to the society." Similarly, agricultural economist Earl Heady has argued that the "development of agriculture has not been left to the free market" (1961:566). In fact, state surveillance and regulation of economic affairs has always been considered legitimate to the extent that it appeared to serve the "public interest." What *has* changed since the mid-1800s is that the state has become increasingly complex and pervasive in scope, and the "public interest" has become equated with expanding industrial capitalism.

Throughout American history, the state has managed to maintain the capitalist order with relatively little forceful suppression of dissidence. Two major alternatives have instead proved successful. First, the government has assumed responsibility for subsidizing the victims of economic dislocation, thus stemming potentially volatile situations of worker insurgency. Strategies and programs for "mopping up" the casualties of capitalist expansion remained state or local affairs through the early part of the 20th century. Following the Great Depression, however, aid for the disenfranchised increasingly emanated from the federal government in the form of New Deal programs, welfare legislation, etc. (Piven and Cloward, 1971).

A second crucial ingredient to social stability has been the ideological legitimation of the capitalist system. Perhaps the best example of the state's support for this cause has been public education. Specifically, the American school system offered an "education" that clearly defined economic concentration after the Civil War as socially beneficial if not inevitable. Historical records show, for example, that "while the nation was in the throes of labor disorder in 1877 . . . the president of the NEA (National Education

Association) said that schools should teach discipline and respect for private property and for the rights of organized industry" (Corwin, 1965:73). Opposition to big business was systematically discouraged by those determining policy in public education. For the skeptical there were doctrines like social Darwinism, which explained that the centralization of wealth and power, "more dominant control, intenser discipline, and stricter methods are but modes of securing more perfect [social] integration. When we perceive this we see that the concentration of wealth is but one feature of a grand step in societal evolution" (Sumner, 1963:151). Hence, public education has imposed a moral obligation on American citizens to abide by the political-economic status quo and so to facilitate the "natural" progression toward capitalist monopolization. We shall return to this point shortly.

In addition to maintaining stability, the state has also provided for capitalist enterprise through (1) political and economic incentives to increase profits and (2) supportive "public" institutions and facilities. In the first case, incentives have ranged from protective tariffs and lenient tax policies to judicial decisions and interpretations that have effectively been used to "frustrate government control of the growing power of monopoly [and] weaken the countervailing power of labor" (Lichtman, 1966:20). Even after the Sherman Anti-Trust Act was instituted, "the Supreme Court wiped out [its] effectiveness against the trusts five years after it was passed. . . . But from 1890 to 1897 the Sherman Act was used twelve times to break labor strikes" (Wasserman, 1972:29). In the agricultural sector, as Mohammadi shows in this volume, legislation in the form of the Homestead Act (1862) and the Pacific Railway Act (1862) was crucial in making possible the domination of farm production by Northern capitalists.

The state has also been instrumental in stimulating industrial capitalism by footing the bill for large-scale projects—on the premise that they served the public good—which may have fostered economic growth and centralization in the private sector but which could not have been operated for profit. The publicly funded irrigation projects in California that primarily benefit huge agribusiness corporations exemplify this type of intervention (Robbins, 1974:34-41). Public education and highway and airport construction and maintenance also qualify as examples. So also must agricultural research be considered in terms of its accommodation to and legitimation of capitalist interests.

AGRICULTURAL RESEARCH AS STATE INTERVENTION

Prior to 1836, state intervention in the American farm economy primarily meant the local regulation of various agricultural practices and products (for example, crop inspection laws). However, in that year the federal government (through the Patent Office) began to collect agricultural statistics, thus sowing the seeds for a national "farm policy" (Sutton, 1977). Since that time, state penetration has essentially meant the support of a pervasive trend toward capitalist agribusiness. State-funded agricultural research has been no exception.

Organized agricultural research in the United States originated through the efforts of dominant political and economic interests toward the end of the 19th century. Specifically, federal support of State Agricultural Experiment Stations via the Hatch Act of 1887 was apparently provided in part as a necessary concession to the growing ranks of economically impoverished American farmers at that time. As Rosenberg explains: "The stations would help the farmers adjust to an increasingly competitive world market, would rationalize and systematize his operations—would provide, that is, a conservative alternative to more radical schemes for adjusting to changed economic... realities" (1976:154). Once created, it became imperative for experiment stations in each state to establish ties with those farmers who would be most receptive to their endeavors. As Rosenberg points out, this was not likely to be the small farmer: "The political needs of the station scientists guaranteed that the educated, adequately capitalized farmer would be their natural ally in the achievement of power. Indeed, the larger the scale of an enterprise, the more likely it was, in general, to find experiment station scientists relevant. Innovation and adoption implied both capital and appropriate values" (1976:170). Moreover, as the incentive for "original research" grew— especially with the Adams Act of 1906—the reciprocity of interests between dominant producers (and, indirectly, their suppliers and buyers) and the experiment station scientists intensified.

The agricultural experiment stations' commitment to agribusiness was not to diminish after the turn of the century. Indeed, Wendell Berry offers persuasive evidence to show how Section 347a of the Smith-Lever Act (1955) first officially acknowledged some farmers as being potentially "too small" or "too unproductive" to facilitate the "inevitable agribusiness revolution." In addition, he contends that "the tragedy of the land-grant acts is that their moral imperative came finally to have nowhere to rest except on the careers of specialists . . . [on] the 'objective' practitioners of the 'science' of agriculture, whose minds had no direction other than that laid out by career necessity and the logic of experimentation" (1977:156-57). From the beginning, then, agricultural research was thoroughly enmeshed in the network of vested interests predicated upon capitalist agricultural development.

Despite this fundamental historical reality, agricultural colleges have understood themselves as being responsive to the needs and concerns of diverse groups of farmers. Jim Hightower argues that this public image is in fact largely an illusion: "Although the land grant college complex was [allegedly] created to . . . reach out to serve the various needs of a broad rural constituency, the system has, in fact, [functioned as] the sidekick and frequent servant of agriculture's industrialized elite" (1973:1). Similarly, Wendell Berry expresses his distress that the "ivory-tower" experts have come to accept agriculture not as a way of life, but as a business, "purely a commercial concern [designed] to provide as much food as quickly and cheaply and with as few manhours as possible, and to be a market for machines and chemicals" (1977:88). The chief task of agricultural research appears to critics like Hightower and Berry to be the development of agriculture as an "efficient," capital-intensive industry that will yield hefty corporate profits.

If agricultural research plays an important role in forcing American farms to fold at a rate of about 2,000 per week (Hightower, 1973:2), it is at least partially because those who are actually engaged in the research process have appeared to accept the change as progressive or inevitable. Claiming adherence to "professional neutrality" or "scientific objectivity," agricultural scientists have insisted on addressing what Habermas terms "technical" rather than "practical" questions: "Technical questions are posed with a view to the rationally goal-directed organization of means and the rational selection of instrumental alternatives, once the goals (values and maxims) are given. Practical questions, on the other hand, are posed with a view to the acceptance or rejection of norms, especially norms for action, the claims to validity of which we can support or oppose with reasons" (1973:3). Specifically, for example, a technical question would be: "*How* do we improve the per acre yield of tobacco for Kentucky farmers?" A practical question might be: "*Ought* we to encourage farm credit if it is going to enhance agribusiness profits at the expense of the small farmer?"

Miliband has observed that "the great majority of academics in [advanced] capitalist countries have found little or no difficulty in reconciling their vocation with support for the 'national purpose,' whatever that purpose might be" (1969:249). Acknowledging capitalist agribusiness to be synonymous with "national purpose," American agricultural scientists appear to conform to Miliband's generalization. To be sure, supportive empirical evidence abounds. The current literature on the relationship between agricultural research and agribusiness indicates that it has been viewed by agricultural scientists as essentially unproblematic (Barham, 1977) or, at most, mildly troublesome (Breimeyer, 1973). Moreover, studies demonstrate that agricultural researchers tend to see virtually no alternatives to an increasingly mechanized (technicized) mode of agricultural production (Fujimoto and Fiske, 1975).

The argument presented thus far in this chapter has suggested that the state has facilitated the accumulation of monopoly profits in the American economy and that agricultural research appears to support this activity. This position is no doubt easier to discuss empirically using a broad rather than narrowly focused frame of reference. Nevertheless, in an attempt to further substantiate what has been said, a critical examination of the position of the Kentucky Agricultural Experiment Station in relation to tobacco production interests in Kentucky is presented. While the complexity of this relationship makes it impossible to offer a comprehensive investigation here, a relatively brief historical sketch should serve our purposes.

AGRICULTURAL RESEARCH AND TOBACCO PRODUCTION IN KENTUCKY

Tobacco was cultivated in Kentucky at least as early as the colonial period and gained prominence as a key export crop by the late 1700s (The Tobacco Institute, 1967). Despite its apparently uneven quality, Kentucky tobacco began to command relatively high prices on the marketplace in the early 1800s. At this time, tobacco processing factories and warehouses became increasingly prevalent in a number of Kentucky communities, including Louisville, Lexington, and Frankfort. What in many counties had been a supplementary crop in the 1700s began to surface as the major cash crop by the first part of the 19th century. Indeed, "in 1821 planters were being advised by farm authorities that tobacco was 'unquestionably the best crop the farmers of Kentucky can at this time raise'" (The Tobacco Institute, 1967:40). It should perhaps be noted that state intervention in tobacco production during this period was confined largely to regulatory activities such as warehouse inspection, quality control of tobacco, and export trade limitations.

Through the mid-1800s, particularly after 1840, there was a continued growth of tobacco processing facilities as well as production of new varieties of tobacco. Markets grew and expanded eastward with the development of railways such as the Louisville and Nashville Railroad. (The Tobacco Institute, 1967). The Civil War caused a slight decline in production when some farmers diverted resources from tobacco to other staple cash crops and livestock. Yet, on the whole, the war was not detrimental to Kentucky's tobacco economy. As Saloutos notes:

> The war . . . caused an important geographical shift in the production of tobacco. From Virginia, the principal producing area before the war, the growing of tobacco shifted westward into the newer fields of Kentucky, Tennessee, North Carolina, and southern Ohio, where land was more fertile and where the effects of the war had not been as disastrous. . . . Thereafter, Kentucky emerged as the foremost tobacco producer in the country (1960:8).

This trend was reflected in the increased tobacco production and acreage devoted to production in the postwar years: between 1866 and 1875 Kentucky tobacco land increased from 142,000 to 320,000 acres (The Tobacco Institute, 1967:45).

As might be expected, tobacco production in Kentucky did not progress without problems for the producers. Stanley Aronowitz suggests that "surplus is the key word in the history of American agriculture" (1974:74). This statement aptly describes the case of tobacco production in Kentucky as early as the late 1800s. Excess production and price instability began to plague tobacco farmers across the state. The economic insecurity which many farmers experienced at the time was largely attributable to the fact that J.B. Duke had established the American Tobacco Company in 1890 and spearheaded the monopolization of tobacco purchasing and manufacturing. That the Duke empire was able to expand to a $25 million enterprise and gain control over the tobacco market was problematic for more than a few Kentucky farmers. "At the start of the twentieth century, some 85 of the 119 counties in Kentucky were dependent on tobacco" (Saloutos, 1960:168).

The economic hardship suffered by tobacco farmers, especially the many hopeful tenant farmers who had flocked to the state to raise tobacco after the Civil War, at the hands of the all-pervasive "Trust" was severe enough to prompt large numbers of farmers to support the establishment of tobacco growers' cooperatives. In the decade between 1900 and 1910 this movement was fraught with problems: legal-legislative barriers initially prohibited combination to raise prices. Policy disputes existed within and among different organizations. Furthermore, the Trust actively attempted to "discourage" farmers from abiding by cooperative agreements. In fact, accounts reveal that corporate hegemony on the part of the American Tobacco Company included the establishment of fake cooperatives, the use of the press to discourage (by threats) banks and merchants from dealing with cooperative members, and even selective "door-to-door" efforts to underbid the cooperatives for the farmers' tobacco. Given these strategies, it is perhaps little wonder that some committed cooperative members took to "night riding," destroying property and in some instances killing those farmers who yielded to the Trust (Saloutos, 1960:167-83).

While these political-economic barriers were variously overcome and limited success was achieved in enhancing tobacco prices (largely because by 1908 the Trust was attempting to cope with federal restraint-of-trade investigations by keeping a low profile), this first tobacco producers' movement went the way of so many other populist movements: "The farmers were left prey to . . . an irrational and unstable marketing system" (Wasserman, 1972:109). In the words of Saloutos, "nothing substantial was achieved" (1960:183).

It is essential to examine the role played by the Kentucky Agricultural Experiment Station (KAES) in the economic events related thus far. Under the Hatch Act the function of each state experiment station was to "aid in acquiring and diffusing among the people of the United States useful and practical information on subjects connected with agriculture, and to promote scientific investigation and experiment respecting the principles and applications of agricultural science . . ." (KAES, 1888:174). A thorough review of tobacco-related articles in the KAES bulletins between 1885, when the institution first began operations, and 1915 reveals that a great deal of emphasis was put on "scientific investigation," and that "useful and practical information" was interpreted to mean exclusively technical, apolitical advice to farmers.

Staffed initially with three chemists, two entomologist-botanists, a "practical horticulturist," a farm superintendent, and a secretary-stenographer, the KAES published its first bulletin in December of 1885. It was entitled "Do Fertilizers Affect the Quality of Tobacco?" Upon discovering "conclusively that the quality and quantity of tobacco can be improved by the application of fertilizer," the researchers explained: "We give these tentative experiments to the public to show the line we *must* follow next year" (KAES, 1885:2; emphasis added). It was precisely this brand of inquiry that the Experiment Station scientists followed for the next three decades or so.

Seven years after Duke initiated his corporate empire, the struggling Kentucky farmer could be comforted by turning to the 1897 research bulletin to find (only) tobacco fertilizer test results or eloquent descriptions of "the life history of the southern tobacco worm" (KAES, 1897:20). In 1907, farmers were informed that "Kentucky is destined to be the center of importance in the tobacco industry of the United States for many years to come" (KAES, 1907:4). The key to a prosperous future, of course, was seen as improvement in yields via better methods of crop handling and construction of properly ventilated barns. In a 1909 bulletin, a brief history of tobacco production in Kentucky from 1775 through the early 1900s completely ignored the American Tobacco Company and the response of many Kentucky farmers to market monopolization. The same bulletin described the growth of the tobacco industry in Kentucky and pictured a series of about a dozen farm implements (such as a seed grader and a disc and drag harrow) that would prove worthwhile capital investments for improving farm operations—hardly welcome news to farmers fighting to stay financially solvent in the depressed tobacco market (KAES, 1909). In short, the KAES disseminated purely technical information to Kentucky farmers at a time when only farmers least threatened by the Trust—landowning, highly capitalized farmers—could begin to appreciate that kind of information.

All this is not to imply that the KAES and the large tobacco interests were linked in a conspiracy against small-scale Kentucky tobacco farmers. What the bulletins do suggest is that the agricultural scientists and those farmers and industrialists committed to agribusiness shared a common interest in "progress" and profits—an interest that ultimately meant opposition to the small farmer's struggle to make a living. The research effort in Kentucky was seen by those involved exclusively as *scientific* activity that enhanced tobacco *production,* and thereby the prosperity of Kentucky. If dozens of fertilizer manufacturers claimed that their product would improve crop yields, then the task of the KAES was to assess the data and announce the results. This was the organization's raison d'etre: to carry out research for the sake of progress. Hence, for example, in one 1910 KAES bulletin there were 120 consecutive pages of tables presenting the names, manufacturers, and characteristics of fertilizers on the market (KAES, 1910: 200-319). Clearly, no concern was voiced about rising profits in the commercial fertilizer industry or the inability of some small farmers to purchase the "approved" brands.

During the 1920s, articles appearing in the KAES bulletins increasingly dealt with economic issues involving tobacco production in Kentucky. This is probably not coincidental. Tobacco prices had generally risen between 1911, when the Supreme Court officially divided the Trust, and 1919 (Cox, 1933:149; USDA, 1927:134). But in 1920 burley prices plummeted from 33 to 13 cents per pound (Nicholls, 1951:215). That there had been a "mutual understanding" among major buyers such as American Tobacco, Lorillard, Reynolds, and Liggett and Meyers to hold out for low prices is hardly a matter of dispute (Nicholls, 1951:207-14). In any case, farmers who had tasted prosperity the year before now faced financial disaster. With farmer unrest mounting and the economic viability of Kentucky's tobacco industry in jeopardy, an organization for tobacco producers was established in Kentucky.

Specifically, by 1920 powerful people in national and local financial and tobacco circles were assembled by *Louisville Courier-Journal and Times* owner Judge Robert W. Bingham to establish the Burley Tobacco Growers Cooperative Association. Helped by the large contributions of businessmen (the president of the National Bank of Kentucky, for instance, pledged a half million dollars), the organization was able to attract over 55,000 members by 1921 (Axton, 1975). After five years or so the Cooperative proved unable to generate price stability. By virtue of its influential founders, however, the organization had drawn public attention to the adverse effects market fluctuation had on Kentucky's tobacco industry. Thus, the economics of tobacco production appears to have become a relevant and legitimate issue for KAES scientists to address.

To the extent that the articles published by the Experiment Station in the 1920s did discuss the economic situation facing Kentucky tobacco farmers, few fundamental questions about the system of private profits, which spelled disaster for increasing numbers of producers, were posed. In fact, they tended to mystify fluctuating market relations as something to which farmers could adapt on an individual basis but could never really change. For example, even articles on cooperative marketing in 1926 and 1928 suggested that these organizations could only be expected to "lessen" price fluctuations from year to year. Moreover, it noted, "Cooperative marketing . . . is not organized production, it is organized selling. The marketing organization necessarily leaves control of production with the individual grower" (KAES, 1926:271; 1928). The farmer's struggle was, then, presented to him as an essentially private affair.

The bulletins indicate, furthermore, that a tobacco farmer's success was to be measured solely by his profit margin and that his profit margin was primarily dependent upon individual willingness and readiness to read the economic signs of the times. For example:

> The production of farm products under present day conditions has become a commercialized enterprise. Profits in farming, as in other lines of production, are in a large degree dependent upon good business management. A farm enterprise, to justify its continuance, should return a net profit above its cost. . . . The farmer's prosperity is more dependent upon *his* ability to control these costs and to expand or change his types of production than upon any other factors (KAES, 1920:137; emphasis added).

The message here is clear: the farmer had to accept the existing system or suffer the consequences. Whether the "pitch" was the use of credit or the purchase of a new farm tool, acceptance of the status quo was always implicitly encouraged. However, given the market situation in the 1920s, it should be noted that such acceptance was by no means a simple task for the tobacco farmer.

Throughout the 1920s tobacco producers had to contend with cigarette manufacturers who reported increasing profits (Tennant, 1950), but who offered prices for burley that remained substantially below the 1919 price (Nicholls, 1951:228). Some manufacturers (for example, American Tobacco and Imperial) refused to deal at all with producers who attempted to sell their tobacco through a cooperative. That American Tobacco, Reynolds, and Liggett and Meyers purchased about 80 percent of the 1929 burley crop is testimony to the limited market available to farmers at the time. Needless to say, matters did not improve with the stock-market crash of 1929. Indeed, tobacco prices took a further disastrous plunge, and by 1931 the burley tobacco farmer could expect to get about 9 cents a pound for his crop (Nicholls, 1951:228).

The market position of some producers was improved with the Agricultural Adjustment Act of 1933. With this act the federal government implemented both a loan system and an acreage allotment program to protect producers from exceedingly low market prices in periods of surplus tobacco production. The federally sponsored price-support program was calculated to guarantee tobacco producers 90 percent of the fair market price of tobacco. The allotment program was designed to encourage producers to cultivate less acreage of tobacco. This encouragement was facilitated in 1936 when Congress passed the Soil Conservation and Domestic Allotment Act providing financial incentives to producers who devoted proportionately less acreage to tobacco, which was recognized and defined as a soil-depleting crop.

The impact of these federal programs soon became clear. Without any restrictions placed on the *amount* of tobacco that could be produced on a given allotment, there was an all-out push for high per-acre productivity. The KAES gained tremendous significance here in that it was very successful in facilitating (via research) increased yields per acre. Yet the solution to this technical problem revitalized the age-old dilemma of surplus production for tobacco farmers! As Axton indicates, "Productivity ran . . . far ahead of acreage restriction" (1975:125). All told, intensified production efforts decreased the bargaining position of smaller tobacco producers because (1) they were less likely to be able to afford the capital investment (fertilizer, machinery, etc.) required for increased yields per acre and (2) they had less ability to absorb the reduction in the support prices due to unsalable tobacco (Axton, 1975:125).

In terms of objective consequences, then, the KAES research enabled better-capitalized farmers to derive disproportionate benefits from the federal price stabilization program. Yet one would never guess this to be the case by glancing through the Experiment Station bulletins in the 1930s. In fact, the KAES's general position regarding disenfranchised farmers in this decade seems to be aptly summed up in a 1938 article, "Science and Agriculture." According to the author: "It is not the province of the experiment station to undertake to interfere with national and international economic policies and government activities. These are problems which can only be solved by the farmer himself" (KAES, 1938:395). This "neutrality" entailed one more instance among many of systematic (though probably unintended) bias toward support for the existing structure of market relations—a structure that historically has tended to favor those tobacco producers and associate concerns that persist as viable profit-making enterprises amid the uncertainty of capitalist market relations.

CONCLUSION

In sum, it is argued that the state plays at least two key roles in capitalist society by virtue of its support for agricultural research. First, by funding this research it provides dominant economic interests involved in agriculture with "the most expensive economic needs of corporate capital . . . [i.e.,] the costs of research, development of new products, new production processes, and so on" (O'Connor, 1974:126-27). Second, the state provides legitimation for capitalist agribusiness via "scientific" research. Marx offers some insight on this point: "Insofar as the State admits the existence of social evils, it attributes them to natural laws against which no human power can prevail, or to private life which is independent of the State" (1964:216). The historical observations made in this chapter regarding agricultural research seem to support this statement. Specifically, agricultural research has been instrumental in legitimating the existing structure of production by (1) ignoring fundamental questions that might challenge or threaten a market system based on agribusiness profit making, or (2) reifying capitalist market relations as an inevitable series of individualistic battles for prosperity. That agricultural scientists have not considered their activities in this light is perhaps not surprising. Structurally, it is participation in this legitimation process that insures the experiment station its bread and butter.

REFERENCES

Aronowitz, Stanley. 1974. *Food, Shelter and the American Dream.* New York: Seabury Press.

Axton, W.F. 1975. *Tobacco and Kentucky.* Lexington: University of Kentucky.

Barham, W.S. 1977. Industry-Institutional Cooperation-Coordination in Horticulture: A Questionnaire. *HortScience* 12:208-10.

Berry, Wendell. 1977. *The Unsettling of America: Culture and Agriculture.* San Francisco: Sierra Club Books.

Best, Michael H., and William E. Connolly. 1976. *The Politicized Economy.* Lexington, Mass. D.C. Heath & Company.

Breimeyer, Harold F. 1973. Public Sector Research and Education and the Agri-Business Complex: Unholy Alliance or Socially Beneficial Partnership? *American Journal of Agricultural Economics* 55, no. 5: 993-96.

Corwin, Ronald G. 1965. *A Sociology of Education.* New York: Appleton-Century-Crofts.

Cox, Reavis. 1933. *Competition in the Tobacco Industry, 1911-1932.* New York: Columbia University Press.

Domhoff, G. William. 1978. *The Powers That Be: Processes of Ruling Class Domination in America.* New York: Vintage Books.

Fujimoto, Isao, and Emmett Fiske. 1975. What Research Gets Done at a Land Grant College: Internal Factors at Work. Davis: Department of Applied Behavioral Science, University of California, mimeo.

Habermas, Jurgen. 1973. *Theory and Practice.* Boston: Beacon Press.

Heady, Earl O. 1961. Public Purpose in Agricultural Research and Education. *Journal of Farm Economics* 43:566-81.

Hightower, Jim. 1973. *Hard Tomatoes, Hard Times.* Cambridge, Mass.: Schenkman Publishing Co.

Kentucky Agricultural Experiment Station. 1885. Do Fertilizers Affect the Quality of Tobacco? Bulletin no. 1:1-2.

————. 1888. Report of the Laws: The Hatch Act. Bulletin no. 16: 174-77.

————. 1897. Notes on Tobacco Worms, from Observations Made in 1896. Bulletin no. 66:6-32.

————. 1907. Tobacco. Bulletin no. 129:3-32.

————. 1909. Tobacco. Bulletin no. 139:5-36.

————. 1910. Commercial Fertilizers. Bulletin no. 153:195-322.

————. 1920. The Cost of Producing Tobacco in Kentucky. Bulletin no. 229:137-90.

————. 1926. Cooperative Marketing and Price Control. Bulletin no. 271:259-79.

————. 1928. The Cooperative Marketing of Tobacco. Bulletin no. 288:273-306.

————. 1938. Science and Agriculture. Bulletin no. 388:386-400.

Lichtman, Richard. 1966. *Toward Community: A Criticism of Contemporary Capitalism.* Santa Barbara, California: Center for the Study of Democratic Institutions.

Marx, Karl. 1964. *Early Writings.* Edited by T.B. Bottomore. New York: McGraw-Hill.

Miliband, Ralph. 1969. *The State in Capitalist Society.* New York: Basic Books.

Nicholls, William H. 1951. *Price Policies in the Tobacco Industry: A Study of "Concerted Action" and Its Social Control, 1911-1950.* Nashville: Vanderbilt University Press.

O'Connor, James. 1974. *The Corporations and the State: Essays on the Theory of Capitalism and Imperialism.* New York: Harper & Row.

Parenti, Michael. 1977. *Democracy for the Few.* 2d ed. New York: St. Martin's Press.

————. 1978. *Power and the Powerless.* New York: St. Martin's Press.

Piven, Frances Fox, and Richard A. Cloward. 1971. *Regulating the Poor: The Functions of Public Welfare.* New York: Vintage Books.

Robbins, William. 1974. *The American Food Scandal: Why You Can't Eat Well on What You Earn.* New York: William Morrow & Co.

Rosenberg, Charles E. 1976. *No Other Gods: On Science and American Social Thought.* Baltimore, Md.: Johns Hopkins University Press.

Saloutos, Theodore. 1960. *Farmer Movements in the South, 1865-1933.* Lincoln: University of Nebraska Press.

Sumner, William Graham. 1963. *Selected Essays of William Graham Sumner: Social Darwinism.* Edited by Stow Persons, Englewood Cliffs, N.J.: Prentice-Hall.

Sutton, S.B. 1977. How the Department Got Its Start. *Horticulture* 55:33-37.

The Tobacco Institute. 1967. *Kentucky and Tobacco: A Chapter in America's Industrial Growth.* Washington, D.C.: The Tobacco Institute.

Tennant, Richard B. 1950. *The American Cigarette Industry: A Study in Economic Analysis and Public Policy.* New Haven, Conn.: Yale University Press.

United States Department of Agriculture. 1927. Prices of Farm Products Received by Producers (Kentucky). Statistical Bulletin no. 16 (June), table 58. Washington, D.C.: Government Printing Office.

Wasserman, Harvey. 1972. *Harvey Wasserman's History of the United States.* New York: Harper & Row.

4

The Establishment of Applied Sciences: Medicine and Agriculture Compared

S. Randi Randolph and Carolyn Sachs

INTRODUCTION

In scientized American society, social values, beliefs, and norms are continually being replaced by technical rules that mystify the social world (Schroyer, 1973). The technocratic world view is embedded in our society to the extent that a commitment to technological progress continues, despite the costs of reproducing social dislocations and isolation. Schroyer explains how science and politics support the technocratic strategy:

> Traditional liberal political practice is being replaced by the *technocratic strategy* in which politics and science are related integrally as the means for a more efficient and effective decision-making process. In this way we have indeed left behind the ideology of cultural values and are now guided only by the "neutral" standards of purposive rational action, or what could be called *instrumental reason* (1973:19; emphasis in original).

Throughout history, pure reason has frequently offered the promise of making the world a better place. However, appeals to pure reason are no longer necessary to solve the problems of the world because many people believe technology will correct them. As Reid (1977) points out, instrumental rationality (or reason) has been institutionalized as both method and metaphysic, which promotes an ahistorical and atheoretical acceptance of the technocratic strategy.

Science both legitimates and provides information for the technocratic strategy. While it is generally conceived of as a method for obtaining knowledge, science also promotes a world view that offers the hope of solving

This chapter was first presented as a paper at the annual meetings of the Midwest Sociological Association, Minneapolis, 1979.

society's problems. Scientific ideology penetrates the Western world view to the extent that science as a belief system can be comprehended in terms of Gramsci's notion of a hegemonic system. Boggs described Gramsci's concept of hegemony as the "permeation throughout civil society . . . of an entire system of values, attitudes, beliefs, morality, that in one way or another supports the established order and ruling class" (1972:98). Hegemony is an organizing principle or world view disseminated by agencies of social control that mystifies events, issues, and power relations (Boggs, 1972). The belief in science as a cure-all is made possible by science's capacity to mystify issues. Scientists as well as nonscientists are often mystified by science.

In order to comprehend the manner in which science operates as a hegemonic system, a historical analysis of the development of two applied sciences, which on the surface appear to share little in common, will be undertaken. The specific concern here is to trace the establishment of science in the fields of medicine and agriculture. By employing medical (or biomedical) science as an analogue for agricultural science, we intend to reveal the permeating influence of the technocratic strategy on the practice of both medicine and agriculture.

Examination of the emergence of the medical and agricultural sciences allows for the possibility of "seeing through" science and of calling into question "taken-for-granted" assumptions that might otherwise be overlooked if the history of either science were examined separately. "Seeing through" science refers to the ability to view science as a product of human activity. Figlio (1977) has noted that the writing of medical history has often served to covertly legitimate the state of medicine as it is, since the history itself is generally written from a positivist stance. Indeed, the history of science has attempted "to place metaphysical disguises upon the faces of process and procedure. After the disguise or mask has been worn for a considerable time, it tends to blend with the face and it becomes extremely difficult to 'see through' it" (Turbayne, 1970:4). He further explains that this process has deluded historians as well as scientists into thinking that they have replaced metaphysics. The inability to "see through" science is the basis of our contention that science has become a hegemonic system.

The change in societal legitimation systems will serve as a background to explain the establishment of science in medicine and agriculture. An analysis will be made of the material structures and ideological forms which both supported and were supported by science so that it became firmly entrenched as a hegemonic system. The capacity of scientific ideology to mystify and legitimate issues will be revealed through a comparison of the impact of the following taken-for-granted assumptions of scientific medicine and agriculture: technical dominance, mechanistic conceptions of the world, and reductionism. Moreover, an attempt will be made to decode various scientific structures and ideologies in order to provide insight into possible alternatives.

THE HISTORICAL FRAMEWORK

The extent to which science as a world view has permeated our society is evidenced by the strict adherence of current medical and agricultural practices to techniques that are legitimated and mystified by science. The question of whether these practices should be based on science is seldom raised. Few remember that the practices of both medicine and agriculture existed thousands of years prior to the inception of science and that they were gradually rationalized and scientized in the Western world during the 19th and 20th centuries. Both emerged directly out of folk knowledge in contrast to the origin of basic sciences in intellectual traditions.

Carlson (1975), in his polemic against the medicine of the 1970s, utilizes a schema that traces the gradual emergence of scientific medicine. He examines five eras in medicine's history that reflect its scientific and social relations with the larger society. This schema has been adopted as the framework of this paper (see Table 4.1). We have enlarged it to include the development of agricultural science, thus providing a framework within which to compare patterns of change. Carlson's analysis does not spell out the specific time spans for each era. In following Carlson, our analysis does not depend on a conception of each era as a discrete stage of development; we utilize the framework simply as a means of comparing general forms of scientific development.

Although the schema adopted in Table 4.1 includes ancient cross-cultural components, our attention here is devoted mainly to the 19th and 20th centuries in American society. We choose to focus our attention on the past and current centuries because it is here that we see a funneling of medical and agricultural practices into what can properly be called science. In this paper we focus on the impact of science on the practitioners of medicine and agriculture—namely, physicians and farmers—and on the scientists who conduct research in each field. In contemporary medicine there is very little distinction between medical practice and medical research, but in the 19th century the practice of health care and research developments were quite separate, especially in the United States (Starr, 1978; Imershein, 1976; Shryock, 1947). Similarly, the practice of agriculture did not form a union with scientific methods until the 19th century. Through a brief overview of the historical development of medicine and agriculture prior to scientization, it is possible to transcend certain taken-for-granted assumptions that are perpetuated through scientific ideology.

Our rationale for the use of this scheme as an overarching structure is that it allows the reader the possibility of generalizing the transformations manifested in medical and agricultural sciences to other applied sciences, such as engineering, business, education, and home economics. Moreover, if, as has been stated, the development of medicine parallels the development of science in general (Imershein, 1976), such a framework allows for insight into the

Table 4.1   Eras of Medicine and Agriculture

MEDICINE

Eras	Dominant world-view toward health	Medical technologies utilized	Prevailing health paradigm
1. Ages of magic	Sickness was not an abnormal human condition but a feature of hard existence. Sickness resulted from imbalances in humans' relationship to their environment	oral record containing lessons about healthy life rituals derived from oral record or empirical observation sacrifice to gods	shamanistic tradition based on balance of two roles: heal patient and community symbolic aspects of healing
2. From shaman to doctor	Idea of body as machine that could be taken apart and reassembled. Understanding of disease processes makes human intervention possible	bloodletting, cauterization, purgatives techniques outside body: surgical, chemical, or electrical	paradigm in flux observation as source of knowledge
3. Public health	Defects in machinery introduced by virulent environment. Ecological concept emerged: balance between species and environment. Skepticism toward therapeutic efficacy	sanitary sewage systems improve water quality	both body and environment treated by mechanics
4. Rise of science of services	Medical care no longer idiosyncratic. Events had causes. Repeatability of intervention produces same results	hospitals to immobilize patient taxonomies of disease precision of techniques	disease-oriented disease malfunction of human machine rise of technical medicine
5. Medicine of today	Disease presumed result of faulty machinery. Body compartmentalized into finer and finer parts	prevalent surgery excessive drug prescriptions unwieldy delivery system marketing by insurance companies	medical care = health

Source: Eras of medicine adapted from Rick J. Carlson, *The End of Medicine* (New York: John Wiley & Sons), pp. 201-12.

Table 4.1 (continued)

AGRICULTURE

Eras	Dominant world view toward food production	Agricultural technologies utilized	Prevailing agricultural paradigm
Magic	Magical practice could help humans control plants, animals, and weather	rain dance ritual sacrifice of plants and animals wood techniques introduced	grow and raise food for community subsistence emphasis on symbols for successful agriculture
Natural husbandry	Plants and animals can be controlled through human intervention	breeding of plants and animals irrigation horse-drawn plows	observation of methods that work
Intensification of agriculture	Ecological approach Fertility of soil decreased, needs improvement	spread of manure crop rotation animals fed in stables	use of practical methods to return fertility
Rise of agricultural science	Solution to food problem can be found through application of science to agriculture	chemical fertilizers bacteriology genetics tractors and mechanical harvesters	commodity-oriented rise of technical agricultural hunger = malfunction of food machine
Agriculture of today	Food problem can be solved through creation of improved technology	control systems feedlots climate control increase in use of machines for harvesting excessive use of fertilizer	possible to control products of agriculture animals and crops must fit machines industrialization of agriculture: the farm as a factory

gradual ascendance of scientific thought in disciplines that have more direct impact upon laypersons than the basic sciences. Through application of scientific ideas and methods to issues of health and food production, the larger society becomes a passive recipient of technocratic strategy based on instrumental rationality. The hegemony of science comes to pervade everyday life.

PRESCIENTIFIC ERAS OF MEDICINE AND AGRICULTURE

Era One

Domestication in neolithic village cultures brought together the two economies of food gatherers and hunters (Mumford, 1966). The entire economy of neolithic villages was intermingled with an array of ancient magical rituals and religious notions tightly bound up with practical achievements. The importance of shamanistic medicine was symbolic. Shamans (or healers) validated their power in tangible effects—such as expectoration of blood-stained down after a "treatment." Group healing ceremonies were events during which the shaman could instruct the larger society about sickness. Gifts and sacrifices were made to gods in order to achieve sufficient domestication of plants so that an adequate supply of food would be available. Sympathetic magic was utilized extensively in this regard. For example, in Ireland a magical symbol for agriculture was the dandelion. It was "associated with the favorite Irish saint, who was a milkmaid, protector of cows, and a successor of a pagan godess" (Mumford, 1966:157).

Neolithic agriculture was practiced by the whole community and yielded a higher level of well-being than had been possible in a food-gathering economy (Mumford, 1966). The village provided protection and continuity for its members; most probably infant mortality and morbidity rates were lowered as a result of the better diets afforded by village life. Domestication of herd animals paralleled seed agriculture, apparently not solely in response to a desire to increase food production, but also for ritualistic reasons. The bull was a frequent sacrifice in these villages, and urine and excrement of sacred animals were often preserved. Mumford speculates that the spread of manure on fields may well have had its origins in ritual. Even the eating of domesticated animals may have, at first, had religious significance as sacrifices equated with the eating and drinking of the body and blood of a god. Eventually, mechanical inventions such as the plow became associated with domesticated animals in ritual, and the combination enhanced seed cultivation. Mumford concludes that rituals and magic were among the factors in early human beings' existence that provided the groundwork for later higher technological development.

During this era, the providers of health care and food for the community appealed to ritual and magic in order to legitimate their practices. The community's views of health and agriculture were similar in that it assumed that human intervention must be accompanied by an awareness of the balance of nature and appeals to spiritual powers. Although both medicine and agriculture relied on similar legitimation systems, divergence emerges in this era. The practitioners of medicine were specific individuals in the community, while all members of the community were practitioners of agriculture. Thus it seems that only select individuals were allowed to possess the knowledge to provide health in times of crisis, whereas the entire population needed to have knowledge of food production practices.

Era Two

Gradually, in Western societies, the shaman was replaced by the physician, and the role of farmer began to be differentiated in the social structure of communities. In the second era of both medicine and agriculture, the notion of human intervention in the name of reason succeeded previous appeals to the supernatural world as the means of improving health and food production. In line with this notion, the conception of the world in mechanistic terms began to emerge. The body was seen as a machine that could be taken apart and reassembled. If disease processes and body responses could be understood, interventions with physical, chemical, or electrical methods were possible (Carlson, 1975). Similarly, the nature of agriculture underwent a fundamental change. Once humans had considered themselves to be part of nature; now they became self-conscious exploiters (White, 1962). Both animal and plant worlds were mechanistically taken apart and reassembled by human intervention to satisfy human needs. Rows served as demarcations during planting; irrigation controlled erratic rainfall; and animals were bred to possess specific characteristics. Human intervention into the ecosystem ultimately took its toll by the latter part of the 18th century. Colonial farmers failed to rotate their crops and neglected the use of animal manure (Gras, 1940; Bidwell and Falconer, 1925.) The farmer in America was faced with declining fertility of the land because natural husbandry often failed to make returns to the land.

Both medical and agricultural practices relied on reason as the legitimation for human intervention. This era marks the emergence of the human belief that nature can be controlled. The physician emerged as the person who was able to control the body, while the farmer attempted to control plants, animals, and water supply. The practitioners of medicine were few, whereas the majority of the population practiced farming. Inadequate attention to the environment resulted in problematic situations for the health of city dwellers and for crop yields of farmers during this period.

Era Three

The 19th century marks a major transformation to environmental emphases in concerns about health and agriculture as the ecosystem showed signs of depletion. The depletion of the ecosystem occurred with industrialization and urbanization. The effects of unsanitary living conditons on morbidity and mortality rates were brought to light by public health authorities who shared a commitment to pietism. As a result, nineteenth-century public health reforms had moral overtones even when science was relied on to produce solutions. As one active reformer remarked, "Disease, like sin, is permitted to exist; but conscience and revelation on the one hand, and reason and science on the other, are the kindred means with which God has armed us against them" (quoted in Rosenberg, 1976:116). Clearly, a belief in scientific solutions was emerging. This period led to the installation of sanitary services and to interventions into the social environment rather than the human body. Such interventions were easily dissociated from medical concern, since physicians traditionally had focused on healing those who were already sick and had been able to ignore the filth and misery surrounding their urban patients. The separation of causes (housing, diet, sanitary sewage, and water systems) from effects (sickness and its symptoms) was reinforced through public health reforms.

A public health movement termed the Popular Health Movement emerged in the 19th century and was comprised of radical activists from several "unorthodox" medical sects that opposed the special privileges of the allopathic aristocracy (Corea, 1977). Much of the emphasis of the Popular Health Movement was on prevention rather than cure. The efforts of Movement advocates resulted in repeal of the medical licensing laws by the mid-1800s. Alongside the public health movement, upheaval and "therapeutic nihilism" were the trademarks of medical practice in the second quarter of the 19th century (Starr, 1978). According to Starr, challenges to the profession came from within and outside of medicine. Various sects of medical practitioners warred against each other, resulting in a loss of professional coherence and public respect. The arcane knowledge of physicians was proclaimed as no more than mystification by the general populace. Physicians themselves were skeptical about the validity of their therapies. .

As a result, changes in the social structure of medical practice took place in the latter part of this era. The spread of hospitals brought medical practitioners into close affiliation with other physicians who shared similar professional interests. Dale and Greer (1978) note that all competing paradigms of medicine were suppressed by the American Medical Association (AMA) in its efforts to promote allopathic medicine as exclusively scientific and solely responsible for scientific progress in medicine. Concurrently with these events, advances in medical science began to be employed in medical practice; and physicians' claims to authority were strengthened (Starr, 1978).

During the third era of agriculture, as in this era of medicine, practices of farming shifted towards an emphasis on the environment. Farmers began applying barnyard manures rather than following earlier habits of dumping the manure in rivers. This practice was based on common sense knowledge rather than scientific evidence of its efficacy. As Whitlock explains, "generations of farmers understood that animal manure and later the refuse of green crops enriched the soil, although they knew nothing of nitrogen" (1965:69). Agricultural scientists meanwhile were pinning their hopes on eventual rationalization of farming through chemistry (Rosenberg, 1976). In the words of one popularizer of this belief, "Every farm should be considered a chemical laboratory and every farmer a practical chemist and philosopher: farming would then be honorable and lucrative" (1976:147). While chemists were searching for scientific procedures to ensure soil fertility by use of fertilizers, the farmers were not yet adopting scientific agriculture.

The damage of unbridled human action to the environment called into question previous practices that had ignored the role of the environment in the health and sustenance of its inhabitants. The stage was set so that physicians and farmers would be receptive to suggestions that they adopt scientific techniques in their endeavors. The inability of humans to control nature through the application of reason did not discredit the belief that humans could control nature, but rather paved the way for an appeal to science as a means of control.

SCIENTIFIC ERAS OF MEDICINE AND AGRICULTURE

After this brief description of the prescientific eras of medicine and agriculture, a more in-depth analysis of the scientific eras will be undertaken. Prior to examining the structural and ideological forms of medical and agricultural science in particular, it seems useful to discuss the nature of science in general with special emphasis on the relationship between pure and applied science. The distinction between pure and applied science is often made within the scientific community; however, a critical discussion of the significance of this distinction is seldom advanced by scientists except in order to justify support for basic research (Greenberg, 1967).

The Nature of Applied and Pure Sciences

In our attempt to illuminate the nature of the scientific enterprise in agriculture and medicine, it should be noted that these two fields are both conceived of as applied sciences, despite the fact that some scientists in both medicine and agriculture do basic research. However, a critique of medical and agricultural sciences requires an examination of the source of the distinction between pure (or basic) and applied sciences, since applied science claims to derive its source of knowledge from pure science. This distinction is rather tenuous, although it is vociferously defended by scientific statespersons (Rosenberg, 1976; Greenberg, 1967). Pure science's mandate is a search for truth or for understanding of a particular phenomenon. Applied science, on the other hand, purports to emphasize the production of results that can be utilized by its clientele. Thus, with applied science, the emphasis is on "what works." In reality, however, the pure scientists' model of science as the mother of technology—a straight-line sequence from knowledge to utility, or pure to applied—finds insufficient confirmation. More appropriately, an intricate network of interactions exists between the pure and applied sciences in which it is virtually impossible to ascertain distinct contributions from either type when examining any artifact or device developed by science (Greenberg, 1967).

The politics of pure science is that the acquisition of knowledge ranks above the application of knowledge (Greenberg, 1967). Furthermore, the notion of pure science assumes that science is performed merely for the sake of pure, theoretical interest. The metaphysical assumption is that there are facts that can be discovered, that exist in and of themselves, apart from any human interest. Habermas reveals the fallacy in supposing science to be "pure" as he quotes Peirce: "Every discovery of science is a gratification of curiosity. But it is not true that pure science is or can be successfully pursued for the sake of gratifying this instinct. . . . Curiosity is their [the theoretical sciences'] motive; but the gratification of curiosity is not their aim" (1971:133).

Pure scientists labor under the false assumption that their work is divorced from the politics and economics of the social world. Moreover, pure scientists have from time to time claimed to have moral and religious motivations for their work. As Rosenberg notes, "As an absolute good, the cause of abstract research justified, even encouraged, any steps which could be seen as furthering scientific investigation—a cause transcending mere personal ambition" (1976:15). Greenberg (1967) discusses the fact that basic scientists' reverence for their profession has resulted in their promulgating quasi-mystical propositions to justify their quests for support. This they have done without questioning whether the expenditure of several hundreds of millions of dollars per year in public funds to enable a handful of physicists to explore the "particle zoo" is fulfilling any conceivable purpose for society. Greenberg is inclined to believe that they think they are pursuing their own curiosities in virtually total disengagement from the society that supports them. In essence, then, pure scientists from their higher professional status denounce concern with the utility of their work and reject the notion that their efforts reflect and affect human values and attitudes while simultaneously accepting and actively pursuing public support of their activities.

In contradistinction, applied scientists do not claim to be pursuing knowledge for its own sake; rather, they emphasize "what works." However, the pure scientists' claim to be above human interest, or neutral, allows the applied scientists to appeal to the ethic of neutrality as well. In the eyes of the scientist, pure science is neutral; and the application of scientific knowledge does not seem to involve value judgments on the part of either the pure or applied scientist. As Schroyer points out, the scientific community "has a naive image of the correspondence of knowledge and nature and is blind to the societal consequences of science" (1973:133).

Applied scientists also adhere to the ethic of progress, which appears to be neutral. However, the ethic of progress has propelled American society to the brink of a scientized technological abyss. The blind faith in the combination of democracy and science as a sure formula for human progress began in Franklin's and Jefferson's epoch (Price, 1965). This ethic has two main tenets concerning progress. First, human beings' desire for material benefits leads to societal support of science and technology while the profit motive concurrently encourages the development of the economy. Second, the advancement of science leads society toward desirable purposes, including political freedom. By appealing to "what works" in terms of achieving progress, the applied scientists do not recognize the pseudoneutrality of their work.

The interest of applied science in progress can also be viewed as an interest in the mastery of nature. DiNorcia (1974) refers to the myth of human mastery that is perpetuated by a focus on successful technological achievements while the negative by-products of these achievements are concealed. Yet the assumption that human control of nature is possible is fallacious. As Berry has noted, total human control is just as impossible at present as it always has been. Total control for the scientists is possible only "within a small and highly simplified enclosure, he simply abandons the rest, leaves it totally out of control" (1977:71).

Furthermore, in Horkheimer's words, "man is stripped of all aims except self-preservation. He tries to transform everything within reach into a means to that end" (1947:101). The applied sciences of agriculture and medicine aim to preserve the human species while simultaneously pursuing a course that devalues the nonhuman world and often the human beings themselves.

Era Four

In the fourth and fifth eras of medicine and agriculture, science began to emerge as a legitimation for the established order and provided support for the acceptance of a technocratic strategy. From the late 19th century to the present, a process occurred in which the assumptions of science became taken for granted. To that extent science has become a hegemonic system. For the purpose of studying hegemonic processes, Livingstone (1976) has suggested utilizing a framework that examines both the material structures and the ideological structures of hegemony. An examination of the material structures in which scientific medicine and agriculture emerge will be undertaken through a brief comparison of the following aspects of the political, social, and economic structure of each science in era four: support of research, the practitioners, and the commodities produced. For era five, we will compare the institutional structures in which the practitioners of medicine and agriculture find themselves and examine the impact of adherence to scientific ideology.

Support of Research in Era Four. In order for science to flourish, it is necessary for scientific research to be supported (Rosenberg, 1976). The sources of funding for research in Western societies are the government and various institutional structures in the private sector. Initial financial support for the development of the medical and agricultural sciences was by no means parallel (Lape, 1955). Through a comparison of the funding sources of scientific research in these two fields, it is possible to comprehend the linkages between science, capitalism, and the state.

Medical research in the last quarter of the 19th century was not accorded the prestige in the United States that Europe, particularly Germany, gave it (Rosenberg, 1976; Shryock, 1947). In the United States, both government and industry were more concerned with the economic utility of science. Hence, while federal aid was available in the 1880s to study the health of farm animals, there was little money available to study human disease (Strickland, 1972). It appears that the lack of government support was due partly to the state of medical science at this time and partly to the fact that the cure of human disease brought little financial return.

Initially, substantial amounts of money for medical research came from foundations established by Rockefeller and Carnegie. The wealth that these industrialists had received from the expansion of industrial capitalism was now being invested in what Greenberg refers to as "socially desirable objectives" (1967:57). Ehrenreich and English (1978) document the dynamics that surrounded the financing of scientific medicine by Rockefeller and Carnegie. In their view, these two "robber barons" were attracted to the scientific claims to impartiality and objectivity as a guide for their philanthropy. "Their charity had to be as seemingly impartial and detached as their money-making had been ruthless" (1978:75). They turned to the experts of philanthropy who had developed "scientific giving" through large centralized agencies. The principles of "scientific medicine" appealed to Rockefeller's philanthropic expert, Gates, who also favored the patrician ideal of "regular" medicine. Carnegie's foundation was approached by the AMA in 1907 for money to finance an "objective" study of medical education. The subsequent Flexner Report was written by a brother of the Rockefeller Institute's director and was responsible for the demise of large numbers of "irregular" medical schools. For the first 40 years of this century, the funding for medical research was primarily private. Moreover, there was little demand from the private sector for government support of medical research until after World War II.

Unlike medicine, the support for agricultural research came predominantly from the government. Agricultural colleges were created by politicians through the Morrill Act. In 1887, the Hatch Act was passed, providing $15,000 per year to each state in order to establish an agricultural experiment station. The question arises of why the state and federal government chose to support agricultural science and not other applied sciences such as medicine. McConnell (1953) notes that the Populist movement in the late 19th century was challenging banking and business interests. The antagonism that farmers felt toward banking and business interests is revealed by the fact that the

Farmers' Alliance, which was a major Populist organization, excluded merchants, bankers, and lawyers from membership (Hadwiger, 1976). Furthermore, Rosenberg explains that the prime motivation of federal and state legislators was to provide financial support for agricultural science in order to placate rural unrest. Science, while claiming to be a neutral force, was supported by the government in response to the Populist threat of farmers. Moreover, the agricultural experiment station "would help farmers adjust to an increasingly competitive world market, would rationalize and systematize his operations—would provide, that is, a conservative alternative to more radical schemes for adjusting to changed economic and demographic realities" (Rosenberg, 1976:154). Several decades later the extension movement was also supported by business and banking interests in order to encourage the more affluent farmers to seek economic efficiency rather than class conflict as a means of improving their lot (McConnell, 1953).

Practitioners in Era Four. Within the capitalist economic system of the United States in the late 19th century, the practitioners of medicine and agriculture were entrepreneurs. Physicians competed for fees, while farmers competed for profits. Physicians were also at this time attempting to eliminate competition by excluding irregular practitioners. The formation of the AMA in 1847 saw its primary order of business as the reform of medical education. Through the vehicle of the Flexner Report, nonscientific practitioners were eliminated and a profession was created (Immershein, 1976; Ehrenreich and English, 1978). According to Ehrenreich and English, the regular medical sects had gained a legal monopoly over the practice of medicine by the end of the 19th century. When the AMA was first established there were three competing paradigms for the practice of medicine: allopathic, homeopathic, and eclectic medicine. The "regular" practitioners—those practicing allopathy—set out to eliminate other types of medical practitioners largely through the establishment of the AMA and a reliance on the legitimacy of science (Immershein, 1976). Thus, in the field of medicine, the entrepreneurs (physicians) attempted to eliminate competition through the creation of a professional society that had its source of legitimacy in the practice of scientific medicine.

The practitioners of agriculture in the late 19th century were farmers. As has been previously discussed, the average farmer was not the initiator of the use of science in agriculture. In order for science to gain credence in agriculture, it was necessary for farmers to adopt scientific farming. However, as Rosenberg explains, the "advocates of scientific research in agriculture had to deal with the assumption—and power—of a laity at once skeptical and credulous" (1976:147). As a strategy to combat this skepticism and credulity, agricultural scientists catered to a clientele that would be convinced of the economic advantages of adopting scientific agriculture: the wealthy and influential rather than the subsistence farmers. These wealthier farmers were persuaded by the scientists that chemistry (i.e., fertilizer) could solve many of their economic problems (Rosenberg, 1976). For instance, the first bulletin published by the experiment station in Kentucky was entitled "Do Fertilizers Affect the Quality of Tobacco?" (KAES, 1885). The title of the bulletin implies that the researcher is aware of the probability that the farmer may be skeptical. Moreover, it speaks to the use of chemicals on the major cash crop of Kentucky. The focus of agricultural research on the problems of the large farmers benefited the practitioners who were already well-to-do while the small farmers became increasingly disadvantaged. In contrast to earlier eras, the nineteenth-century economic system did not require that the bulk of the population know how to produce food. The farmers who did not create large marketable harvests began to lose both their land and their credibility.

In both medicine and agriculture, science provided a "neutral" justification for eliminating various practitioners. As nonscientific views of medicine and agriculture were discredited, those physicians and farmers who were involved in scientific practices enjoyed economic prosperity. Nonscientific practitioners eventually began to disappear. The provision of health care and food, which were, respectively, the explicitly stated goals of medical and agricultural practitioners, were transformed by scientific farming and medicine into the pursuit of economic rewards.

A distinct difference is apparent between the practitioners of medicine and agriculture. Medical practitioners—physicians—were trained as scientists themselves, while agricultural practitioners—farmers—were generally not trained to be scientists. As a result, the communication of scientific information between researchers and practitioners took different forms for the two sciences. The AMA's Council on Medical Education sought ways to standardize training for physicians. Together with the Carnegie Foundation, they asked Abraham Flexner to investigate the quality of medical education and to prepare guidelines for reform. His recommendations in 1910 were for close affiliation between researchers developing medical technology and students of medicine. This fostered the growth of large medical centers associated with universities. However, farmers had no such professional organization to implement training programs. Researchers in experiment stations desired to be free of the process of dealing directly with farmers (Busch, 1978). In 1914, the Smith-Lever Act created the Cooperative Extension Service, which placed an intermediary between the agricultural researchers and the farmer. The role of the extension service was to translate technical reports into language that could be understood by practitioners. The farmer is removed from the researcher by language and setting, unlike the physicians, who undergo scientific training in the same setting as the medical researcher. Compared to the physician, the farmer in the early 1900s was placed in a dependent and submissive relationship to the scientific establishment. The authority of the extension agent over the farmer is somewhat comparable to the authority of the physician over the patient. The inability of the patient or the farmer to communicate on a scientific level with authorities encourages the mystification of scientific knowledge.

Commodification in Era Four. In order to comprehend the development of medical and agricultural sciences within the U.S. economic system, it is helpful to examine the organization of the system in relation to commodities. Marx describes the significance of the division of a product into a useful thing and a thing of value: "This division of a product into a useful thing and a thing of value becomes practically important, only when exchange has acquired such an extension that useful articles are produced for the purpose of being exhanged, and their character as values has therefore to be taken into account, beforehand, during production" (1967:73).

As Dale and Greer (1978) have argued, health has become a commodity that can be bought and sold. The commodification of health is made possible through the medical sciences' ability to define disease as a discrete entity and their claim to control disease through scientific intervention. The patient may purchase health from physicians in the form of cures or treatment for specific diseases. Taussig (1980) likewise illustrates the process of rationalization that has taken place in medicine and shows how this process is analogous to rationalization of commodity production. He argues that the power to heal has been converted into the power to control. As the exchange value of health care becomes the primary motive of medical science, the actual use value of healing becomes distorted.

Similarly, the appeal that agricultural scientists made to farmers was primarily one of economic promise. Agricultural science thereby supported the needs of the expanding capitalism of the United States. In order for agriculture to contribute to the market system, food must be a commodity as well as an object of utility (i.e., something to be eaten). The move from subsistence farming to commodity-producing, scientific agriculture involves an increased emphasis on the exchange value of food. Agricultural scientists working within the capitalist system focus their research on cash crops. Busch (1978) notes that early American agricultural research focused on grains, which were the major export crops of the United States. Farmers, who hoped to support themselves, relied on the production of crops for their exchange rather than use value. Those farmers who were organized into various commodity groups—e.g., corn growers, dairy operators, and tobacco growers (Rosenberg, 1976; Smith, 1965)—were most accepting of and accepted by the agricultural sciences. They also became a major source of political support of the agricultural sciences. Thus, in this era, both agriculture and medicine were increasingly being practiced and researched for the purpose of making a profit.

Era Five

The institutional structures that supported the establishment of medical and agricultural science during the late 19th and early 20th centuries in the United States have been altered with the rise of corporate capitalism. Scientific knowledge, through the creation of complex technologies, contributed to greater capital accumulation and centralization in medicine and agriculture.

A very brief description of the current institutional structures of scientific medicine and agriculture is provided. Then we examine various underlying assumptions of the scientific practice of both medicine and agriculture in an attempt to reveal the impact of adherence to scientific ideology upon the practice of medicine and agriculture. The aspects of scientific ideology we compare are technical dominance, the mechanistic metaphor, and reductionism.

Institutional Structures of Era Five. At the present time, medical complexes are the primary setting for health care in the United States. Within these complexes the physicians' control of health care has been increasingly limited. Dale and Greer (1978) document the increasing intrusion of drug companies, insurance companies, equipment supply companies, and medical information systems suppliers into medicine.

What is occuring in medicine is also evident in agriculture. The farmer is increasingly dependent on the nonfarm sector of the economy. The funds for agricultural research now come primarily from industry (Wilke and Sprague, 1967). Agribusiness corporations have instituted systems of contract farming that are patterned after the contracts of industry (Higbee, 1963). Agricultural research has become increasingly related to the marketing and processing of food. For instance, much of the research funded by marketing interests in California has been channeled into "such programs as the prune sorter, strawberry stemmer, raisin stemmer, fig harvester, melon harvester, lettuce harvester, and the development of a number of fruit tree harvesters" (Fujimoto and Kopper, 1975:8). Such research favors large farmers, while more and more small farmers must leave their farms because they are unable to compete. As the report of the National Research Council (1972) suggested, the changes that have occurred in marketing and processing of food have outpaced our knowledge of the nutritive value of such food and the nutritional needs of the population.

Both medical and agricultural science provide support for the corporate economy. The hegemonic function of science seems to have reached its peak in the corporate state. The belief that medical and agricultural science will solve the nation's health and food problems prevails. The promise of science prevades our society despite the fact that a significant proportion of the population is not able to purchase the results of scientific medicine and agriculture except through welfare programs administered by the federal government: Medicare, Medicaid, and food stamps. Science pervades our society to the extent that many ideological assumptions of science have become taken for granted. Let us examine three of them.

SCIENTIFIC IDEOLOGY IN ERA FIVE

Technical Dominance

The assumption that nature can be conquered by people guides the scientific community. Yet the desire to conquer nature can be pursued only if there is a denial of nature in people. To neutralize human nature, a radical separation in science between the active human subject and the passive object (nature) has been nurtured (DiNorcia, 1974).

Both medical and agricultural scientists strive to control processes of nature. Agricultural researchers separate the plant and animal species and operate on them as passive objects. Medicine, by separating illness from the human organism, considers individuals as passive objects. Disease receives active attention while individuals must passively submit to the ministrations made possible by biomedical science and technology. This has directed attention of physicians away from what might be called a period of "being ill" in a person's life toward disease entities entirely separable from patients themselves (Figlio, 1977). Noble Prize winner Frances Crick captured the essence of this phenomenon when he wrote: "Americans have a peculiar illusion that life is a disease which has to be cured. . . . Everyone gets unpleasant diseases and everyone dies at one time. I guess they are trying to make life safe for senility" (quoted in Carlson, 1975:62). Even entry into and exit from existence have been brought under the control of scientific technology.

The miracle of birth has piqued the curiosity of medical practitioners for centuries (Figlio, 1977; Rich, 1977). The birth process was wrested from midwives' quiet assistance and waiting for natural processes to move at their own pace when male practitioners introduced technological intervention by means of the forceps (Arms, 1975). Each time this intervention effort worked, individuals became convinced of their power to influence and control nature. Now, biomedical scientists have even gone so far as to place the very act of conception under their domination through the technique of cloning.

Ausubel et al. note their apprehensions about the prospect of cloning: "Theoretically, cloning could be used to turn out people on an assembly-line basis, each a carbon copy of the other, with whatever traits seemed desirable to whoever controlled the process" (1974:32). The words of prominent geneticists indicate their desire for eugenics programs in the absence of moral or ethical debate. The possibility of creating new forms of life has become a reality with the technology of recombinant DNA (Clark et al., 1980). Although scientists have not yet made such a claim, our interpersonal participation in the perpetuation of our own species may no longer be necessary. We will become passive recipients of scientific technology if genetic engineering continues on its present course. Scientists or politicians will determine what types of people will be born, how many will be produced, and whether parentage will become obsolete. The conception process has been transformed into a potential source of social control by science.

Examination of changes in animal science also reveals an increasing attempt by the scientists to achieve control. As Byerly notes, controlled environments for livestock production are becoming more common. "Increased ability to control genetic change through genetic engineering, reproduction through hormonal manipulation, animal behavior through early environment, and viral disease through immunization seems immanent"

(1976:272). However, he also notes that the increased control of environments may lead to abnormal behavior patterns in animals. These changes in animal behavior were unexpected by scientists who conceived of animals *merely* as food for humans rather than as living organisms.

Berry points to a model of the farm-of-the-future, which was presented in an issue of the *American Farmer,* as an exemplar of the agricultural scientists' attempt to gain total control of nature. On this farm, of A.D. 2076, livestock will be housed in a 15-story building, while all crops will be grown under plastic covers with climate control. As Berry states: "Confronted with the living substance of farming—the complexly, even mysteriously interrelated lives on which it depends, from the microorganisms in the soil to the human consumers—the agriculture specialist can think only of subjecting it to total control, of turning it into a machine" (1977:70).

Questions that are seldom raised within either medicine or agriculture are: (1) how can that which is nonhuman be controlled without at the same time controlling human beings, and (2) if life itself can be controlled, who will control those who are in control?

Conception of Nature in Mechanistic Terms

The mechanistic conception of nature pervades our everyday thinking and reveals itself clearly in the way in which biomedical and agricultural researchers approach their subject matter. The metaphor that people and nature can be looked at *as if* they are machines has so permeated both sciences that its status as a metaphor has been concealed. Both people and nature are conceived of and dealt with *as* machines. As the conception of human beings as machines grew further and further from its metaphoric origins, the emphasis of science has been placed increasingly on substituting "the made" for "the given," since "the made" is perceived to be under greater human control (Shepard, 1977). The objects of scientific interest have become those that are machinelike—that is, objects that are recurrent, manipulable, measurable, and predictable (DiNorcia, 1974)—and those that allow for possible technological control.

Scientists have abandoned the task of understanding human beings as wholes and have turned to a seriatim approach to acquiring understanding of the various aspects of human nature (Carlson, 1975). Once medicine made the Cartesian division between mental and physical states, it chose the body and its functions as its object of interest. Thereafter, only a minor change in the metaphysic was needed to begin to equate the workings of human organisms with the functions of machines. The metaphor has transcended itself in medicine.

Disease was, and is, viewed as a breakdown in the machine. For modern medicine, disease is a discrete entity and is isolated and categorized as something foreign to the body (Taussig, 1980). Diseased organs or limbs can be excised or amputated from a malfunctioning machine and replaced with manufactured prosthetic devices or substitute organs. "The givens" of human beings can now be discarded for "the made." This mechanistic approach is clear in Hamilton's proud display of 32 manufactured replacements for the body which "can make life more bearable for those who have been deprived of perfect anatomy by genetic chance, disease, or accident" (1973:281). In addition, overt misuse of the machine metaphor is apparent when he writes, "Death comes only when the machine is switched off. . . . Technology can do battle with death itself" (1973:280). Now life, as "a given," is under the control of "the made." Scant attention is given to quality of life considerations; the passage from life to death is reduced to a simple bipolar state: on or off. The wholeness of the person is powerless against technology and the mechanistic mentality that encourages and permits the sustaining of lifelike functions in otherwise nonliving individuals—nonliving in the sense of comatose, as was Karen Ann Quinlan.

As far as agriculture is concerned, once again we refer to Berry, who provides an excellent critique of the implications of mechanistic thinking by agricultural scientists:

> If animals are regarded as machines, they are confined in pens remote from the source of their food, where their excrement becomes instead of a fertilizer, first a "waste" and then a pollutant.
> If plants are regarded as machines, we wind up with huge monocultures, productive of elaborate ecological mischiefs . . . more susceptible to pests and diseases than mixed cultures and . . . more dependent on chemicals.
> If the soil is regarded as a machine, then its life, its involvement in living systems and cycles, must perforce be ignored. It must be treated as a dead, inert chemical mass.
> If people are regarded as machines, they must be regarded as replaceable by other machines. They are regarded, in other words, as dispensable. Their place on the farm is safe only as long as they are mechanically necessary (Berry, 1977:90).

Proof of these assertions proliferates in the agricultural trade journals.

As Breimeyer (1977) has noted, there is a continual trend to separate livestock and poultry production from feed-crop production. Many corn farmers in the Midwest do not raise hogs or cattle. The separation of animals from their source of food has resulted in the need for new innovations in animal feeding. An article in *Feedstuffs* reports that the American Alfalfa Corporation has developed alfalfa cubes for feeding horses (*Feedstuffs*, 1978). This cube, smaller than a bar of soap, is designed to replace baled hay. Thus, rather than the farmer raising hay and horses simultaneously, the feed corporation will supply processed food to the horses. The separation of livestock from their food source is most extreme in large-scale confinement operations. Tuten refers to the attitude of an industrial promoter of large-scale farming: "Large scale hog farming is not well suited to a general crop and livestock farm. . . . The name of the game is keeping the production unit operating at full tilt. This is a factory. Things must be done daily. You can't take a few months off to plant and harvest a crop" (1978:10). Even if so inclined, the farmer is no longer able to grow enough grain to supply the large-scale operation. Promoters of large-scale hog farming are attempting to achieve total control environments similar to those already developed for broiler chickens.

Gordon Millar (1969), vice-president in charge of research at John Deere, offers an example of treating plants as if they were machines in his discussion of the task of designing a complete mechanical lettuce system. In order to pick lettuce with a mechanical harvester, lettuce must be grown that will fit the machine. Millar describes how the irregularity of lettuce seed germination and non-uniform plants impede the task of harvesting lettuce. Therefore, agricultural scientists have eliminated the variability in seed germination and have developed seeds that grow into heads of lettuce of uniform size and shape. The process of making lettuce plants predictable is possible through the use of hormone treatments and controlled microclimate for lettuce seedlings. Plants are manipulated by humans to be compatible with machines, so that machines are able to harvest the plants.

In sum, in both medicine and agriculture, the world is dealt with *as* a machine. The assumption is that any problem which arises can be solved with a scientific answer—bigger and better techniques and technology. The body becomes a machine with replaceable parts. Diseases require the services of mechanics. Meanwhile, the farm becomes a factory with inputs increasingly coming from off-farm sources, and the farmer becomes a mechanic for machinelike animals and plants.

The Impact of Reductionism

Reductionism in scientific research is described by DiNorcia as an " 'analytic' method typical in modern research, which as specialization, isolates the sciences from each other, from interaction with practical knowledge of the same subject matter, from the natural environment of their subject matter, and from related social and human problems" (1974:92). Price (1965) states that the endless specialization in science, a by-product of the ethos of progress, has yielded greater and greater abstraction of reality by scientists. As scientists turn their attention away from specific objects of everyday life experience to abstract qualities that can be measured and related to other abstractions, the implications of their research are seldom critically examined.

A manifestation of the reductionistic impact of Cartesian thinking is seen in the proliferation of specialties among medical practitioners who often engage in clinical research. Illich (1976) has documented the fact that the number of specialties recognized by the American Medical Association has doubled between 1960 and 75, with no hint of a halt in this tendency. Half of the practicing American physicians are specialists in one of 60 categories. Needless to say, costs of medical treatment increase with specialization, as do potential risks to patients (e.g., drug toxicity) attendant with this increasingly fragmented approach to medical treatment. Moreover, patients have a difficult time finding a physician who will take continuing responsibility for the whole patient. These highly skilled specialists are frequently ill suited to meet the typical demands of first-contact situations (Fuchs, 1974). Carlson (1975) calls this a "reductionistic drift" that depersonalizes medical-care interactions. Human values and needs are neglected as practitioners treat smaller and smaller parts of the body and research focuses on the minute and esoteric interests of specialists.

Berry states that "it is ... absurd to approach the subject of health piecemeal with a departmentalized band of specialists. ... A medical doctor uninterested in nutrition, in agriculture, in the wholeness of mind and spirit is as absurd as a farmer who is uninterested in health" (1977:103). Ridiculous as its sounds, the current situation finds researchers in agricultural science knowing more about animal nutrition than they do about what comprises a healthy diet for the people who raise the animals.

The specializations within agricultural science are comparable to the fragmentations of subject matter in medicine. Agricultural colleges are divided into departments of agricultural economics, agronomy, animal sciences, horticulture, entomology, and rural sociology, among others. Within each of these various departments, the researchers are considerably more specialized. For example, agronomists are soil scientists or crop scientists. The crop scientist spends most of his/her time researching a particular crop. Animal scientists specialize by specific fields as well as by animals. For example, the *Journal of Animal Science* is divided into the following six sections: applied animal science, breeding and genetics, meat science and muscle biology, nonruminant nutrition, ruminant nutrition, and physiology and endocrinology. Although this type of specialization and research appears quite natural within the structure of modern science, the problems with such an approach are numerous.

Specialization results in an inability to comprehend the whole picture. Due to the process of specialization, a problem and its solution often become two distinct problems. The separation of animal science from crop science is mirrored in the separate locations of animals and crops. Cattle in feedlots are separated from their source of food, while at the same time they produce manure that must be disposed of, since there are no field crops nearby that can easily utilize the manure. Manure disposal, which could have helped resolve the fertilization problem, becomes a problem itself.

Another offshoot of reductionistic orientations is a focus on only what can be perceived immediately, as reflected in both medical and agricultural scientists' approach to their subject matters. Medical practice in the United States is erroneously assumed to be not only the most scientifically pure but the *only* way medicine can be practiced (Imershein, 1976; Starr, 1978; Carlson, 1975). Medical practitioners and clinical scientists remain within the premises of allopathic theory and the disease-oriented approach wherein "signs" (e.g., elevated blood pressure) and "symptoms" (e.g., headaches) are the acceptable types of diagnostic information utilized in treatment and research efforts. Alternative forms of treatment such as acupuncture are received, at the very least, with skepticism by physicians and drug researchers, even though their efficacy has been demonstrated. Medicine has largely refused to acknowledge the healing power of non-Western methods at variance with allopathy (i.e., yoga, meditation, psychic surgery) because they do not fit conventional categories. The tendency to discredit anomalous findings is legendary in scientific circles (Greenberg, 1967; Kuhn, 1970) and reflects both the vested interests of the scientists involved and their disinclination, or inability, to assume a reflective critique of their scientific endeavors.

This narrow focus is well exemplified by a recent result of tomato-production research in California. Since tomatoes are bought on the basis of weight, agricultural scientists attempted to aid growers by increasing the weight of their tomatoes. The following description of the result of this endeavor reveals how focusing only on what can be immediately perceived does not necessarily improve agricultural products.

> In the 10-year period from 1960-1970, it is estimated that tomatoes purchased by processors in California increased on the average of ten per cent by weight—but the increase was all in water content. The result of the efforts of agricultural extension centers and growers was, therefore, to add 450,000 extra tons of water to the 1971 California tomato pack—and to the processor's cost of purchase. But 86 percent of tomatoes are processed, the first step of which is water removal. Hence, this "productivity improvement" is then added to the cost of drying out the tomatoes for ketchup, tomato sauce and other processed tomato products. The consumer ultimately paid the water bill (National Commission on Productivity, 1973:18).

However, in their comment on this occurrence, the Commission suggests that the agricultural experiment station only reacted to what they believed the correct standards to be. The pursuit of science that focuses only on increased productivity discourages scientists from forming alternative approaches to agriculture and seeing the full consequences of their reductionistic strategies.

FUTURE VISIONS

Cultural Critique

Both medical and agricultural sciences have transformed practical questions into technical questions that can be solved only by the application of science and technology. However, adherence to instrumental rationality has led to contradictions in the sciences of medicine and agriculture, some of which are spelled out in the current paper. Schroyer suggests that "our contemporary crisis is rooted precisely in this growing trend toward an instrumental rationalization" (1973:220). Even though all facets of everyday life are increasingly penetrated by technological rationality, many of the problems of everyday life continue to be unresolved. Fortunately, it appears that the domination of instrumental rationality produces the conditions for its own demise. At the same time that scientific ideology permeates society, cracks in the hegemonic system of science appear.

The Frankfurt school has critiqued the category of rationality from the standpoint that human subjectivity is being eliminated. The crisis of the hegemonic system of science is revealed as science's failure to satisfy individual needs as experienced at the personal level. This personal experience can be "raised to the level of interpersonal expressions and become the beginning point of political organization"(Schroyer, 1973:223). Resistance to cultural hegemony currently exists on the interpersonal level. Black liberation, the women's movement, and farmers' revolts form a silhouette of future visions.

Transcending the Mechanistic Metaphor: A Viable Alternative

In regard to medicine, there have been protests against "professionalized" medical care, advances in the popularity of folk medicine, and the organization of self-help clinics. No less important are the trends toward holistic and humanistic medicine within the ranks of physicians themselves. Certain scientific agricultural practices are being questioned by a growing number of critics (Berry, 1977; Hightower, 1973). Practices such as organic farming and an increase in backyard gardens offer alternatives to commercial energy-intensive approaches to growing food. Farmers throughout history, and again in 1980, have staged protests against agricultural policies that do not benefit them. Agricultural scientists have begun to address energy and environmental problems, as fuel shortages and chemical poisoning have become public issues. Even the USDA has recently published a lengthy report on problems associated with changes in the structure of agriculture (USDA, 1979). Although the cracks are small, they are not insignificant.

Schroyer (1973) notes that a dialectic of domination and emancipation exists. For the purpose of attempting emancipation from the dominance of the scientific world view, it is necessary to "extend the possibilities of thinking" (Samples, 1976:90) beyond the hegemony of science, which has defined the manner in which the issues of health care and food production are approached; the use of an alternative metaphor is proposed.

The metaphor by which science, and the technocratic society in general, is guided was formulated by Descartes as follows: "I have hitherto described this earth, and generally the whole visible world, as if it were merely a machine"(as quoted in Turbayne, 1962:13). It is suggested that the mechanistic metaphor that has guided and legitimated technocratic society has been misused. It has been forgotten that the machine metaphor is actually a *pretense* that the world is a machine. In Turbayne's words, "the victim not only has a special view of the world but regards it as the only view, or rather, he confuses a special view of the world with the world" (1962:27). The limitations of the machine metaphor are becoming apparent.

Metaphors may either limit or extend ways of thinking. At one time the machine metaphor extended ways of thinking; now it has come to limit alternative ways of knowing. Our purpose here is to extend current thinking by transcending the machine metaphor, thereby providing an avenue to emancipation. An emancipatory metaphor was sought, following Samples's suggestion that "the ultimate source of metaphor is nature. Just as nature includes humans, so too does it provide the wellsprings of metaphor" (1976:108). A number of metaphors from nature could be and have been used to conceptualize the world (i.e., equilibrium, differentiation, cycles, fertility, birth; see Berry, 1977; Schumacher, 1975; Houston, 1978; Samples, 1976). Durkheim (1933) and Parsons (1977) utilized the metaphor of biological differentiation to explain the movement from primitive society to modern industrial society. Nisbet (1969) notes that the metaphor of growth has governed Western thought concerning mankind and culture throughout history. Naturalistic metaphors, in and of themselves, offer no guarantee of emancipation from the dominant ideology. For instance, as Nisbet recognizes, the metaphor of growth has outlived its usefulness in the sense that it has become reified. Recent suggestions of metaphors for future visions include "fertility" (Houston, 1978) and "birth" (Samples, 1976). Following Houston and Samples, we attempt to stretch the boundaries of contemporary thought not through advocacy of a return to a "simplistic" conception of the world, but rather with a new way to deal with complexities. Of the many metaphors that could be drawn from nature, we have chosen to use the lunar cycle. This choice results from an inspection of the underlying assumptions of newly emerging health and agricultural practices.

The dominant scientific ideology appears to be in the process of transition. This period of flux yields alternative approaches to traditional institutional practices. In Kuhn's (1970) terms, we have a paradigm in crisis. Alternative health and agricultural practices seem to either explicitly or implicitly utilize various metaphoric qualities derived from nature, signaling a dissatisfaction with the existing structures of mechanistic thought. We will attempt to show how approaching health and agricultural concerns in terms of a naturalistic rather than a mechanistic metaphor provides the possibility of emancipatory visions for the future. Others have recognized the metaphoric quality of the lunar cycle. Addison used poetry to tell of the metaphoric qualities of the lunar cycle:

> Soon as the evening shades prevail,
> The moon takes up the wondrous tale,
> And mightily to the listening earth
> Repeats the story of her birth
> (as quoted in West, 1978).

Ideas of change, growth and birth are embodied in the moon's appearance.

In order to utilize the metaphor of the lunar cycle, the following three characteristics of the lunar cycle will be translated into approaches to health and agriculture: transformation, limits of growth, and interconnectedness. The process of tranformation may be aptly juxtaposed to the notion of technical dominance. The degree of illumination of the moon is forever in flux. Rather than assuming that as a natural process it can be brought under total control, a focus on the lunar cycle reveals a process of continual tranformation. Rich (1977) suggests that there is a distinction between power over others and transforming power. Medical scientists, who consciously or unconsciously subscribe to the machine metaphor, attempt to gain power (control) over natural processes such as birth and death. An acceptance of the process of transformation would be translated in the field of health care into such practices as midwifery and the deprofessionalization of healing. With these practices, we see a switch from medical practitioner's attempts to control their patients to a role of assisting the patient in transforming processes. Utilization of the lunar cycle as a metaphor with which to approach agriculture is complicated by the fact that, for centuries, farmers have relied on the cycles of the moon as a guide in planting their crops. The lunar cycle, in addition to being a metaphor with which to understand agriculture, may have an actual effect on the growth of plants. Nevertheless, if food were thought of in terms of its transformative power in relation to the human body rather than primarily in terms of a cash crop, food would be grown with a significant emphasis on its nutritive value.

Upon observing the lunar cycle, it is apparent that the moon waxes and wanes. Once the full moon occurs, the moon begins to wane. The illuminated surface of the moon that is visible from earth reaches a limit and then begins to recede. In health care, there is a limit to the size to which hospitals can grow and continue to provide adequate and satisfying health care services. Recognition of the limits of growth is implemented through the establishment of community health clinics. With an ideology that recognizes the limits of growth, agriculture is practiced on small-scale or subsistence farms. In both medicine and agriculture, the notion that bigger and better technology will improve control over nature is replaced with the idea that technology is useful if it is used on an appropriate scale.

The moon *appears* to change shape throughout the lunar cycle. However, in an attempt to understand the relationship among the moon, the sun, and the earth, it is necessary to go beyond the *appearance* of the moon. What seems to be a change in the shape of the moon is then revealed to be the changing relationship (or interconnectedness) of celestial bodies. An awareness of the interconnections of the human body, both internally and with its external environment, would suggest an approach to health problems that goes beyond the diagnosis of which particular part of the body is unhealthy to an attempt to

Figure 4.1 An Alternative Image

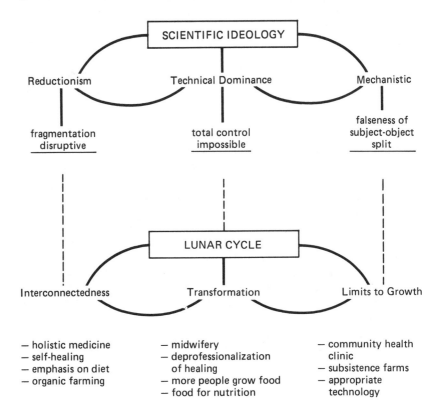

reveal the interconnections of a particular disease with other aspects of the person's body and life. Rather than attempting to remove a diseased part through surgery, self-healing practices or changes in diet might be suggested. In agriculture, an understanding of the interconnections between soils, plants, and animals would manifest itself in such practices as organic farming.

Political and economic interests are intricately interconnected with both our modern social ailments and the "cures" utilized by scientific agriculture and medicine. An awareness of how social, environmental, and occupational hazards contribute to modern illnesses, such as cancer, heart disease, and respiratory disease, becomes potentially emancipatory. Similarly, an understanding of the factors behind production of nutritively poor, chemically toxic, and mechanically produced foodstuffs provides the possibility for inventive political action. The complexity of human existence can be understood as a matrix of interconnections.

The potential for emancipation exists in metaphor. As Nisbet states, "Achieving the impossible is what metaphor is all about" (1969:241). As he recognizes, metaphor can either limit or expand our thinking. We invite the reader to make a comparison of scientific ideology with thinking based on the lunar-cycle metaphor (Figure 4.1). We contend that this metaphor provides the possibility of expanding and extending our individual and collective consciousness at the same time that it permits a critique of science.

REFERENCES

Arms, Suzanne. 1975. *Immaculate Deception*. Boston: Houghton Mifflin Company.
Ausubel, Frederick; Jon Beckwith; and Kaaren Janssen. 1974. The Politics of Genetic Engineering: Who Decides Who's Defective. *Psychology Today* (June):30-43.
Berry, Wendell. 1977. *The Unsettling of America: Culture and Agriculture*. San Francisco: Sierra Club Books.
Bidwell, Persey W., and John I. Falconer. 1925. *History of Agriculture in the Northern United States 1620-1860*. Washington, D.C.: Carnegie Institute of Washington.
Boggs, Carl, Jr. 1972. Gramsci's prison notebooks. *Socialist Revolution* 11:79-118.
Breimeyer, Harold F. 1977. The Changing American Farm. *Annals of the Academy of Political and Social Science* 429:12-22.
Busch, Lawrence. 1978. The Agricultural Sciences and Rural/Urban Development. Paper presented at the 9th World Congress of Sociology, Uppsala, Sweden.
Byerly, T.C. 1976. Changes in Animal Science. In *Two Centuries of American Agriculture*, ed. Vivian Wiser, pp. 258-74. Washington, D.C.: Agricultural History Society.
Carlson, Rick, Jr. 1975. *The End of Medicine*. New York: John Wiley & Sons.
Clark, Matt; Sharon Begley; and Mary Hager. 1980. The Miracles of Spliced Genes. *Newsweek*, March 17:62-71.
Corea, Gena. 1977. *The Hidden Malpractice: How American Medicine Mistreats Women*. New York: Jove Publications.
Dale, Christopher, and L. Sue Greer. 1978. Non-decisionmaking and Resistance to Health Care Innovation: A Community Case Study. Paper presented at the Third Annual Sociological Convention, Notre Dame University.
DiNorcia, Vincent. 1974. From Critical Theory to Critical Ecology. *Telos* 22 (Winter): 85-95.
Durkheim, Emile. 1933. *The Division of Labor in Society*. New York: Free Press.
Ehrenreich, Barbara, and Deidre English. 1978. *For Her Own Good: 150 Years of the Experts' Advice to Women*. New York: Anchor Books.
Feedstuffs. 1978. Alfalfa Cube for Horses Offered. *Feedstuffs* 50, no. 13 (March 27): 17.
Figlio, Karl. 1977. The Historiography of Scientific Medicine: An Invitation to the Human Sciences. *Comparative Studies in Society and History* 19:262-86.
Fuchs, Victor R. 1974. *Who Shall Live? Health, Economics and Social Choice*. New York: Basic Books.
Fujimoto, Isao, and William Kopper. 1975. Outside Influences on What Research Gets Done at a Land Grant School: Impact of Marketing Orders. Paper presented at the Rural Sociological Society meetings, San Francisco.
Gras, Norman Scott Brien. 1940. *A History of Agriculture in Europe and America*. New York: F.S. Crofts and Company.
Greenberg, Daniel S. 1967. *The Politics of Pure Science*. New York: New American Library.
Habermas, Jurgen. 1971. *Knowledge and Human Interests*. Translated by Jeremy J. Shapiro. Boston: Beacon Press.

Hadwiger, Don F. 1976. Farmers in politics. In *Two Centuries of American Agriculture,* ed. Vivian Wiser. Washington, D.C.: Agricultural History Society.

Hamilton, David. 1973. *Technology, Man and the Environment.* London: Faber & Faber.

Higbee, Edward. 1963. *Farms and Farmers in an Urban Age.* New York: Twentieth Century Fund.

Hightower, Jim. 1973. *Hard Tomatoes, Hard Times.* Cambridge, Mass.: Schenkman Publishing Co.

Horkheimer, Max. 1947. *The Eclipse of Reason.* New York: Oxford University Press.

Houston, Jean. 1978. Re-seeding America: The American Psyche as a Garden of Delights. *Journal of Humanistic Psychology* 18:5-22.

Illich, Ivan. 1976. *Medical Nemesis: The Expropriation of Health.* New York: Bantam Books.

Imershein, Allen W. 1976. The Medicalizing of American Society. Unpublished manuscript. Tallahassee, Fla.: Florida State University, Institute for Social Research.

Kentucky Agricultural Experiment Station. 1885. Do Fertilizers Affect the Quality of Tobacco? Bulletin no. 1:1-2.

Kuhn, Thomas S. 1970. *The Structure of Scientific Revolutions.* Chicago: University of Chicago Press.

Lape, Esther Everett. 1955. *American Medical Research: In Principle and Practice.* Vol. 1, *Medical Research: a Midcentury Survey.* Boston: Little, Brown & Company.

Livingstone, David W. 1976. On Hegemony in Corporate Capitalist States. *Sociological Inquiry* 46:235-50.

Marx, Karl. 1967. *Capital.* Edited by Friedrich Engels. New York: International Publishers Co.

McConnell, Grant. 1953. *The Decline of Agrarian Democracy.* Berkeley: University of California Press.

Millar, G.H. 1969. Objectives of Industrial Research. In *Research with a Mission,* pp. 39-48. Madison: American Society of Agronomy Special Publication no. 14.

Mumford, Lewis. 1966. *The Myth of the Machine: Technics and Human Development.* New York: Harcourt Brace Jovanovich.

National Commission on Productivity. 1973. *Productivity in the Food Industry.* Washington, D.C.: U.S. Government Printing Office.

National Research Council. 1972. Report of the Committee on Research Advisory to the U.S. Department of Agriculture (Pound Report). Washington, D.C.: National Technical Information Service, PB 213 338.

Nisbet, Robert A. 1969. *Social Change and History.* London: Oxford University Press.

Parsons, Talcott. 1977. *The Evolution of Societies.* Edited by Jackson Toby. Englewood Cliffs, N.J.: Prentice-Hall.

Price, Don K. 1965. *The Scientific Estate.* Cambridge, Mass.: Belknap Press.

Reid, Herbert G. 1977. Critical Phenomenology and the Dialectical Foundations of Social Change. *Dialectical Anthropology* 2:107-30.

Rich, Adrienne. 1977. *Of Woman Born.* New York: W.W. Norton & Co.

Rosenberg, Charles E. 1976. *No Other Gods: On Science and American Social Thought.* Baltimore, Md.: Johns Hopkins University Press.

Samples, Bob. 1976. *The Metaphoric Mind.* Reading, Mass.: Addison-Wesley Publishing Co.

Schroyer, Trent. 1973. *The Critique of Domination: The Origins and Development of Critical Theory.* Boston: Beacon Press.

Schumacher, Ernst F. 1975. *Small Is Beautiful.* New York: Harper & Row.

Shepard, Paul. 1977. Place in American Culture. *North American Review* (Fall):22-32.

Shryock, Richard H. 1947. *American Medical Research: Past and Present.* New York: Commonwealth Fund.

Smith, J. Allan. 1965. *The College of Agriculture and Home Economics: University of Kentucky.* Lexington: University of Kentucky.

Starr, Paul. 1978. Medicine and the Waning of Professional Sovereignty. *Daedalus* 107:175-93.

Strickland, Stephen P. 1972. *Politics, Science and Dread Disease: A Short History of United States Medical Research Policy.* Cambridge: Harvard University Press.

Taussig, Michael T. 1980. Reification and the Consciousness of the Patient. *Social Science and Medicine* 14B:3-13.

Turbayne, Colin Murray. 1970. *The Myth of Metaphor.* Columbia, S.C.: University of South Carolina Press.

Tuten, Robert T. 1978. Pennfield Plans Porkers for Pennsylvania. *Feed Management* 29:9-25.

United States Department of Agriculture. 1979. *Structure Issues of American Agriculture.* Washington, D.C.: Economics, Cooperatives, and Statistics Service, AER-438.

West, Susan. 1978. Mean History in a Seashell. *Science News* 114:426-28.

White, Lynn, Jr. 1962. *Medieval Technology and Social Change.* London: Oxford University Press.

Whitlock, Ralph. 1965. *A Short History of Farming in Britain.* London: John Baker.

Wilke, H.L., and H.B. Sprague. 1967. Agricultural Research and Development by the Private Sector. *Agricultural Science Review,* 5 no. 3:1-8.

5

Sources of Influence on Problem Choice in the Agricultural Sciences: The New Atlantis Revisited

Lawrence Busch and William B. Lacy

Recently, a number of observers have suggested the importance of examining the agricultural sciences as social products that themselves shape the broad processes of social and economic development (Picou, Wells, and Nyberg, 1978; Busch, 1980; Saint and Coward, 1977). A key issue (Haberer, 1969; Forbes, 1978; Krige, 1978) is the degree to which the products of the agricultural sciences are predetermined by the subject matter (internalist view) or shaped by various social and economic pressures on scientists and research organizations (externalist view).

In this chapter we examine recent developments of the internalist and externalist positions. Next, we present evidence of internal and external factors that influence research problem choices in the agricultural sciences utilizing data obtained through in-depth interviews with scientists. From this evidence we conclude that the internal and external positions are best viewed as distinct strata in the life-worlds of practicing scientists. Thus, for each discipline and institution, science is a continuous process of multilevel negotiation regarding the sources of influence on research.

This chapter was originally presented in somewhat different form at the Rural Sociological Society meetings; Burlington, Vermont, 1979. The authors' names are listed in alphabetical order to indicate equal contribution to this research. The investigation reported in this paper (no. 79-14-45) is a portion of a project of the Kentucky Agricultural Experiment Station and is published with the approval of the director. The authors wish to thank C. Milton Coughenour and Laura Lacy for their comments on an earlier draft.

THE INTERNAL/EXTERNAL DEBATE

In brief, the internalist position (e.g., Merton, 1973; Zuckerman, 1978) centers around the assertion that science as an institution is autonomous.* Science has attempted to maintain its "prejudice against prejudice" (Gadamer, 1976) and thereby create knowledge free from the constraints of human circumstance. Therefore, attempts to direct science are regarded as likely to produce erroneous results and retard scientific progress. The infamous Lysenko affair is frequently pointed to as an example of what can happen if the state attempts to direct scientific affairs (cf. Lewontin, 1976). Moreover, internalists generally argue that decisions regarding project funding should be essentially the result of a peer review and that science policy should be made by scientists.

Internalists thus tacitly accept the Popperian, positivist, or even Cartesian philosophy of science as an accurate account of the way in which science is carried out (Krohn, 1977; Radnitzky, 1973; Krige, 1978). Indeed, Zuckerman (1978:86) suggests that the imperfect correspondence in theory choice between "scientists' actual behavior" and "espistemological prescriptions of how they should behave" may be due to a gap between philosophical concepts and sociological indicators. Of particular importance is the fundamental acceptance by internalists of the notion that scientific knowledge is an accurate representation of the world and hence not subject to external manipulation. The larger social world only impinges upon the world of science to the extent that nonscientific criteria are employed in the allocation of research funds (Nicholson, 1977) or in terms of the broadest features of research policy. When exceptions are found, they are taken as merely evidence of paradigmatic immaturity (Nagi and Corwin, 1972:17). Furthermore, Gaston argues that "whether a country organizes scientific research policy and funding in a *centralized* or *decentralized* method is the interesting dimension of the social organization of science" (1978:34; emphasis in original).

As a result, internalists have tended to focus on "science indicators" such as citation analysis and other similarly "objective" techniques for the study of science. Both the larger sociocultural context of scientific research and the day-to-day affairs of worklife in the laboratory have received short shrift from the internalists. Indeed, the internalists have followed Merton's dictum to develop "theories of the middle range" (Merton, 1967:39-72). Predictably, internalists tend to conclude that "recent case studies of problem choice often find in actual practice what methodological dogma has long maintained

---

*Even the definitions of "internalist" and "externalist" are colored by one's stance in the debate. For example, Kuhn (1977:32n.) sees Merton and apparently himself as externalists in that they take account of institutional factors. He admits the issue is confused, however. In contrast, Feyerabend (1975) clearly demonstrates that social and economic factors external to science enter not only into the organization of science but into the very products of scientific endeavor. From this perspective, both Kuhn and Merton must be seen as internalists.

should be the case: that scientists define some problems as pertinent, and others as uninteresting or even illegitimate, primarily on the basis of theoretical commitments and other assumption structures" (Zuckerman, 1978:74).

In contrast, the externalist position asserts that science is a product of the social environment in which it is constructed. Taken in an extreme vulgar Marxist perspective, science becomes merely an ideology concealing class interests. Scientific claims to objective knowledge are summarily dismissed as a mode for concealing the interests of the ruling class.

A more recent, and more moderate, formulation of the externalist position (e.g., Whitley, 1974) asserts that science shares some but not all of the ideologies prevailing at a given time. Furthermore, it is claimed that these ideologies affect a number of aspects of science. Initially, it is maintained that they affect the form of organization of the sciences. The division of the sciences into broad areas as well as into specific disciplines is said to be societally determined. Second, it is suggested that economic factors play a role in determining topics to be studied and their degree of depth. On this point externalists are in substantial agreement with the internalists. However, unlike the internalists, they are less likely to see the state as a neutral body mediating the interests of various classes. Instead, they view the state as the servant of class interests.

It is also argued by externalists that societies determine the degree of permissible deviation from the established scientific orthodoxy. This is particularly true in the social sciences, where an official doctrine may prevail. The natural sciences, however, are by no means immune to this kind of manipulation (e.g., Rosenberg, 1976).

The reward system is also influenced by the society at large. While the publication of journal articles and the receipt of scientific awards may accurately reflect the reward system among elite scientists, for peripheral scientists both nationally and internationally other factors such as pay, prestige, ability to attract research funds, and administrative scope may be of far greater importance (cf. Gaston, 1978; Cole, 1979). This is particularly true in Third World countries, where competition at the international level is unlikely, if not impossible.

It has also been suggested by externalists that the society at large influences, and often determines, the choice of research problems. As one biologist has noted (Luria, 1973) the increase in scale of scientific activities and the advent of direct state funding has tended to turn researchers into opportunistic research entrepreneurs. This, of course, is of little consequence if it serves merely to direct scientists to research problems of social significance. However, "the danger is that a man may come to identify his scholarly function with the specific goals of certain sectors of the power structure" (Luria, 1973:80).

Finally, externalists argue that the society may influence scientific research styles or orientations. This is perhaps the most subtle way in which scientific research is influenced by the larger society. Specifically, it is argued that the society at large may influence the kinds of instrumentation used, the locale in which research takes place, the design of experiments, and the way in which research results are reported. Thus, this assertion squarely challenges the internalist position that science is essentially an autonomous activity.

Recently, the externalist position has been bolstered by two related changes: the decline of the belief in the unity of science and the changing role of science in society. Let us briefly examine each of these societal shifts.

The Problem of the Unity of Science.

The internalist position initially rested on the argument that, while one may have the choice of doing X or Y first, eventually both must be accomplished. This was asserted since both were considered to be part of the same jigsaw puzzle. In contrast it was argued that external influences would result in the development of "things" that were not part of the scientific puzzle at all.

The work of Thomas Kuhn (1970) dealt a near-death blow to the unity-of-science position. This was ironic in that Kuhn's work was a contribution to the *International Encyclopedia of Unified Science*. Perfect knowledge was no longer viewed by either scientists or those who study them as a valid goal. Then it could be argued that knowledge of some aspects of the world must be developed at the expense of knowledge of others. Therefore, decisions as to what to study must be social decisions rather than technical ones. Moreover, one might have many separate sciences developing out of many cultural systems. This is not to say that such scientists would produce contradictory results, but that they would deal with different aspects of the natural world. The complementarity principle (Bohr, 1958) has lent strong support to this position.

The Changing Role of Science in Society

It is naive to assume that the structure of modern science should have remained unchanged since its inception in the seventeenth century. Indeed, as the status of science changed from that of a peripheral to that of a core institution, so has its role in the larger society. During its formational period, science was clearly seen as a liberating force, freeing humanity from the bonds of tradition and scholasticism (Rose and Rose, 1976:9; Maritain, 1944:10). The scientist, like the atomistic entrepreneur of Adam Smith, could do what he felt was important, with full assurance that his work would be guided by an equally invisible hand (of nature?), and become a building block in the cathedral of knowledge.

With the development of the corporate form, science became corporatized. Big science began to replace little science (DeSolla Price, 1963). Rather than master craftsmen in their own house, scientists became employees—first of the state (in agriculture) and later of industry. Moreover, this shift from Cartesian to Baconian science (Haberer, 1969; Busch, 1980) changed the very nature of scientific knowledge: "the dominant mode of production of scientific knowledge has become that of knowledge-as-commodity, as a marketable good with a cash value" (Rose and Rose, 1976:15). In addition, the copyright and patent laws have contributed to this commodification.

The increase in scale, the development of scientific "manpower," and the increasing centrality of the relations between science and the state have forced the external influences on science into the public (and the sociological) eye. Not only have questions of science policy become definable and, hence, a matter for debate, but both the state and the large corporations have become increasingly reliant upon scientific knowledge to bolster and maintain their legitimacy (Habermas, 1970). Moreover, both have taken an active role in the suppression of research results they considered contrary to their interests.

These developments have lent support to the externalist position precisely because science itself, as an institution, has become less autonomous. Indeed, it can now be argued that the claim of scientific autonomy has become an ideology designed (much like the calls by the giant corporations for "free enterprise") to further certain class interests.

In the study reported below, we examine the agricultural sciences, a group of disciplines that are self-consciously "applied" or "mission-oriented." Since their inception, these sciences have been committed to *both* the accumulation of scientific knowledge and the pragmatic goal of increased productivity (Rosenberg, 1977). As such, they permit us to study a wider spectrum of scientists (Zuckerman, 1978:82) and to examine the internalist and externalist positions in a new light.

METHOD

To examine the internal and external sources of influence on the choice of problems within the agricultural sciences, this study utilized (1) semistructured interviews with key informants in the agricultural sciences, and (2) a review of the publications, research and educational guidelines, and formal organization of each agricultural science.

The interviews contributed a particularly heterogeneous and detailed qualitative information base for assessing the processes and factors related to scientists' problem formulation and the development of the disciplines. Through open-ended questions the interview schedule provided information on the informant's current research, influential people in problem formulation, theoretical orientations, important considerations for choice of research, topics within their discipline that are consistently avoided and the reasons, the perceived audience of their research, sources of financial support, frequency of interaction with scientists and nonscientists, and opinions about each scientist's discipline

The population for the interviews consisted of agricultural scientists in the fields of agronomy (crop and soil), animal sciences, biochemistry, genetics, horticulture, agricultural engineering, entomology, forestry, nutrition, food science, and phytopathology. In-depth interviews were conducted at several agricultural colleges. The sampling strategy employed was that of theoretical sampling or dimensional sampling (Glaser and Strauss, 1967; Arnold, 1975). This strategy maximizes the variation in responses by interviewing persons in very different circumstances. Each interview was concluded by asking the scientist to suggest another interviewee whose research orientation differed from that of the respondent. The use of this approach permits the development of theoretical categories that reflect the views of the practitioners of each of the various sciences, rather than those of the investigators. Therefore, these scientists are best viewed as key informants rather than as respondents in a representative sample. Moreover, theoretical sampling substantially reduces the risk that an artificial symbolic framework is thrust upon the research population. This may be of particular importance in the case of scientists, who are actively engaged in the social production of symbolic frameworks.

In addition to the interviews, for the years 1974-77 the major journals of each discipline were reviewed for state-of-the-discipline articles, critical letters, and other materials that reflect the official views, research emphasis, and publication orientation of the disciplines. This provided an additional comparison of the various disciplines, an analysis of the relationship of discipline organizations and policies to research problem formulation, and of the extent of the influence of internal and external sources on development in each discipline.

Below, we present some of the internal and external influences upon the agricultural sciences. This conceptual framework emerged from our interviews and appears to reflect the several finite provinces of meaning through which both scientists and observers make sense of their life-worlds. Occasionally, a particular phenomenon may emerge as both "internal" and "external." Thus, for example, instrumentation may be alternatively viewed as objects-at-hand for use in the resolution of a disciplinary problem or as products of a larger social universe that impinge upon the researcher. However, the interpretation given will reflect the context within which the phenomenon is discussed.

INTERNAL INFLUENCES ON AGRICULTURAL SCIENCE

A number of internal factors appear to impinge upon the agricultural research process (Fujimoto and Fiske, 1975). A particularly important factor is the process by which researchers are socialized (Fisch, 1977). It is commonly assumed that one's major professor in graduate school is the most influential person in shaping research orientations. While this may be true, major professors during postdoctoral training, sabbatical leaves, colleagues on the job, and even one's graduate students are frequently mentioned as key figures in research choice and paradigm orientation.

A leading agricultural scientist suggested that the sources of influence may be changing in his field of biochemistry. The selection of a particular paradigm and research direction for the oldest generations of biochemists was probably influenced greatly by the doctoral training. In contrast, the middle-aged biochemists as well as many of the younger scientists have often been most influenced by their professors during their postdoctoral training and sabbatical fellowships. Finally, many younger scientists may be additionally influenced by their first job due to limited opportunities to pursue prior training and tight job markets. A young entomologist reflected that both of his current research projects were suggested by department colleagues. A pomologist and an entomologist both indicated that the introduction to their respective research fields was the result of a teaching assignment. And a plant pathologist observed that opportunity and colleagues at one's first job are the primary bases on which members in his field choose a particular orientation. While it is true that the particular people most influential in shaping a scientist's research may vary from individual to individual and from discipline to discipline, key fellow scientists remain important internal sources of influence.

Of equal importance, is the content and structure of the training and educational programs. For example, the paradigm and research orientations of biochemists are greatly shaped by training in either chemistry or biology. Similarly, agricultural meterologists will employ either physiological or statistical approaches in their attempt to explain the impact of climatological factors on plant growth. In addition, the approach taken in many of the plant and animal sciences may depend upon whether one has received training in genetics or physiology. In entomology it has been suggested that the broader research questions involving genetics and microbiology were avoided because training in the basic sciences has not been emphasized. Indeed, even the development of an ecological perspective appears to be the result of undergraduate and graduate training.

A third area of internal influence is the development of new theoretical orientations and methods. For example, several important research topics have emerged in biochemistry with recent theoretical and methodological breakthroughs, such as determination of the structure of DNA and techniques for preparing recombinant DNA. Similarly, in the field of animal science, the development of new methods for assessing the nutritive content of feeds opened up substantial new avenues for research. New discoveries and improved methodology have also played an important role in research development and direction in pomology. Here theoretical work on alternative pathways for electron transfer in mitochondria, as well as the availability of inhibitors and techniques for maintaining mitochondria in test tubes, have led to increased emphasis in this area.

A fourth area of internal influence involves the historical development of the various agricultural disciplines. In the 19th century, it was assumed that agricultural science would develop as an essentially unitary discipline.

Pioneering work in the field made it clear that some specialization would be necessary. Over the years the number of specialties has multiplied, and each of these specialties has established relatively well-defined discipline boundaries. For example, food scientists, concerned with the development of new food products, have tended to differentiate themselves from nutritionists. Indeed, in many universities nutrition departments are found in the medical college rather than in agriculture. In the area of biochemistry certain areas, such as organic analysis, were avoided, not only because they were difficult and complex but also because of the strong historical chemistry tradition in the field. Another example of historical influences on scientific development is found in both entomology and meteorology, which developed as disciplines overlapping with agriculture. As a result, in both fields scientists have seen a need to differentiate agricultural questions from other research questions. Historically the two disciplines have addressed this issue in quite different ways. In meterology, where those concerned with agriculture are in a distinct minority, a separate professional society for agricultural meteorology has been established. In contrast, in entomology, where agriculturally oriented "economic entomologists" far outnumber their liberal-arts counterparts, only one professional society now exists. However, a review of the heated debate in their newsletter suggests that the legitimacy of certain research topics is still a real issue to a substantial minority of the membership. Thus, disciplinary boundaries affect the range of acceptable research topics.

Finally, at most universities the existing set of disciplinary boundaries is maintained through the process of faculty selection. Generally, positions are earmarked for use by a particular department, and the determination of specialties within the disciplines is the result of departmental decisions. As a result, redirection of research and introduction of major reconceptualizations within a discipline occur only slowly.

EXTERNAL INFLUENCES ON AGRICULTURAL SCIENCE

Perhaps the most important external influence upon agricultural research orientations are the various commodity groups that financially and politically support selected aspects of agricultural research (Fujimoto and Kopper, 1975). While the intimate relationships between agricultural research and the various commodity groups have been documented elsewhere (Hightower, 1973; Berry, 1977), it is important to note that commodity groups are in large part the creature of agricultural research. It is only with the development of highly integrated, monocultural "factory farms" that commodity specific interests have emerged. The more powerful commodity groups tend to encourage research on key problems that broadly affect the "industry" they represent. They may do so either by supplying funds directly to selected researchers or through successful lobbying for additional state or federal money. Several state commodity boards have received research and operating funds through taxes on the sale of the commodities, which they direct to specific researchers and research topics. For example, several agricultural scientists whom we interviewed reported receiving funding from state crop advisory boards. One leading pomologist reports that he has received funding from a number of such boards, including the walnut, prune, and olive growers. A young entomologist reports that his research on biological control of particular pests had been suggested by his chairman following contact with the fig growers in the state. A senior plant pathologist indicates that his current research was prompted by feedback from farm advisers and that he has received funding from several growers' associations. A viticulturist suggests the extremely important role that the state wine institute and wine advisory board have played in research in the field. Alternatively, in Kentucky many agricultural scientists at the Tobacco and Health Research Institute are directly supported by research funds and research facilities from a tax on Kentucky tobacco sales. Such funding procedures and lobbying efforts may, and frequently do, influence research choices and result in the overfunding of selected research topics (Evenson and Kislev, 1975). Moreover, the lack of commodity groups in such areas as soil science and rural sociology often puts them at a noticeable disadvantage.

Social movements have also had an impact on the agricultural sciences (e.g., Cheney, 1974). In recent years the environmental movement has forced the reexamination of the role of chemicals in the production of agricultural goods. With the publication of Rachel Carson's *Silent Spring* (1962) the reductionist character of much agricultural science was substantially challenged. In the intervening years substantial shifts in perspective have occurred. In entomology, integrated pest management employing a variety of insect-control techniques is fast replacing the almost exclusive reliance on chemical means. Similar changes have occurred within phytopathology, but not without resistance. In the words of one researcher, "we must make chemical control a science rather than a necessary evil. We must think of chemicals as disease management tools rather than as killers" (Gilpatrick, 1976:2). In addition, at least one new field, environmental toxicology, has emerged as a result of environmentalist pressure.

Consumer concerns and increasing societal interest in health and nutrition have also influenced the development of the agricultural sciences. According to a young food scientist, the consumer movement has led to growing interest in research on food additives and food microbiology. Concurrently, researchers in the field of nutrition report that these same concerns are altering research directions in their discipline.

Agricultural research, unlike more basic research in the biological sciences, must also take into account the changing economic context. For example, until recently virtually all soil classification schemes were created with production uses in mind. Moreover, American agricultural research has tended to focus upon an economic environment in which land, energy, and capital were relatively cheap, while labor was expensive. As a result, American farmers today tend to have highly capitalized—some would say overcapitalized—operations that employ very heavy doses of fossil-fuel energy. With the formation of the OPEC oil cartel and the subsequent rise in oil prices the balance among these factors of production was sharply altered. As a result, energy efficiency has become an issue not only within agricultural engineering, but also in other agricultural disciplines. For example, the increase in shipping costs due to energy consumption has sparked increased interest in postharvest physiology. If postharvest deterioration of plant material can be stopped or reduced, then slower and more energy-efficient means of transportation may be employed to haul food to market. For example, a pomologist reports that the two hottest topics are (1) research on alternative pathways of electron transfer in mitochondria that may increase shelf life of produce, and (2) research on certain plant hormones that are fundamental in controlling ripening. Similarly, in entomology the increased cost of insecticides (which are nearly always petroleum-based) has sparked the search for alternative means of pest control. Finally, the high cost of nitrogen fertilizers has spurred research in biochemistry on nitrogen fixation.

Economics may also play an important role in determining topics that are underresearched, ignored, or avoided. For example, an animal scientist reports that his research on double muscling in cattle had been virtually an ignored topic of research, primarily because it was economically unimportant to breeders. A parallel example was provided by a pomologist who indicates that research on the behavior of gases in cellular and intercellular context had become an ignored topic 30 or 40 years ago. At that time the economic goals of this research had been achieved without a concomitant understanding of the molecular process. Nevertheless the field was abandoned as a major area of inquiry.

Finally, the very pursuit of research may be dependent on funding and therefore susceptible to changing economic conditions. Many of the scientists in the study replied that they would not have pursued their current research if funds had been unavailable. Others went further to indicate that the costs of retooling a laboratory and the current investments in laboratory equipment made radical changes in research problems and orientation difficult, if not impossible. Indeed, an agricultural-college dean recently lamented that he was the caretaker of a 10-year-old building filled with obsolete equipment.

The mission orientation of the agricultural sciences also makes them responsive to government regulation of agriculture. For example, recent federal government regulations have required the licensing of pest-control applicators and the restriction of a large variety of insecticides. This has led to a marked shift away from chemical control within entomology. Indeed, one department chairman reports that he has experienced substantial difficulty in finding new faculty with an interest in that area. Moreover, those few who still remain staunch advocates of chemical control as the solution to all insect problems have earned several derisive epithets, among them "spray and pray," and "the kill 'em and count 'em boys."

Government regulations may also spur heated disciplinary debates (Dunlap, 1978). For example, environmental and health regulations generally provide legal definitions of toxicity. Many chemicals, such as saccharin, are toxic or carcinogenic only in enormous dosages. Current regulations require that toxicity be established by feeding rats with the largest dosage that is physiologically tolerable. While some toxicologists find this definition acceptable, informants tell us that most members of the discipline find it "unrealistic." The problem, however, is the lack of an objective standard that might be substituted.

Government may also influence the direction of research through requests for proposals (RFPs). For example, problems involved with dam maintenance led to the formulation of an RFP for the development of foliage tolerant to flooding in reservoirs. Similarly, the Department of Agriculture has recently announced a competitive grants program to develop basic knowledge relating to photosynthesis. The impact of government granting policies appears to be strongest at the most prestigious institutions. While many experiment stations rely primarily on noncompetitive "Hatch" and state funds, a few receive relatively large proportions of their annual budgets in the form of competitive grants. Indeed, in these institutions faculty salaries may be paid with "soft money," thereby encouraging faculty to become "research entrepreneurs."

The development of instrumentation also plays a significant role in directing the path of scientific investigation. For example, the development of tissue-culture techniques and equipment has opened wide new areas of research in a number of disciplines and promises completely to bypass well-established genetic techniques. Similarly, in pomology gas chromatography has stimulated increased research on plant hormones. In addition, mass spectrometers and lasers have affected research problem choices in biochemistry. Finally, nearly all science has been altered by the development of computerized data banks. For example, most agronomists and horticulturists previously recorded the results of field experiments in notebooks. Computerized data storage has permitted both the more rapid processing of statistical data and the use of more complex statistical techniques.

On the other hand, a lack of instrumentation may prohibit the development of certain aspects of a discipline. For example, agricultural meteorologists have generally avoided the measurement of wind and turbulence within plant canopies. In large part this has been due to the high cost and maintenance difficulties involved in instrumentation.

In recent years demands for accountability have had the effect of encouraging research administrators to measure productivity more accurately. Counting publications in refereed journals has proven to be a convenient, if flawed, measure of output. A number of our informants suggest that the pressure for publication has narrowed markedly the choice of research problem. For example, little work has been done in forestry on the breeding of improved varieties due to the extremely long reproductive cycles of trees (Minckler, 1976:215). Similarly, few long-term (five to ten years) studies of wildlife in their natural habitat have been performed for much the same reasons.

At the personal level, a scientist's nonscientific interests and life-style may influence research-problem choices. One scientist who engages in a great deal of field research indicates that his choices were greatly influenced by his desire to work outdoors. A plant pathologist also reports that his research interests developed from his hobby of ornamental horticulture.

IMPLICATIONS

It is clear from the evidence presented above that the agricultural sciences are influenced in a wide variety of ways by a diverse array of factors. Some of these influences are clearly disciplinary in character, while others represent the demands of outside interests.

It appears that decisions as to what problem to study are clearly not assignable to any single cause or group of causes. Indeed, virtually all research decisions appear to be the result of a complex of influences, some of which are internal to science and some of which are external. Recent research by Knorr and Knorr (1978) on the construction of a scientific research paper appears to further support this conclusion. They note that organizational structure, interpersonal relationships, and methodological difficulties, as well as the scientist's disciplinary concerns, shaped the problem choice.

The very fact that one cannot study everything simultaneously opens the way for external factors to influence the research process. Some of these have been illuminated by our research. It appears likely that the visibility of these external influences varies inversely with the distance of the research problem from the world of everyday life. Hence, subatomic particle physics, whose objects are, by definition, not directly sensible, is little influenced by the larger social world. In contrast, the social and medical sciences, whose objects are of daily concern to all, are susceptible to a whole range of outside influences. Within the agricultural sciences a similar pattern exists. In this research, basic scientists were more likely to respond with an internalist perspective regarding the influences on their disciplines and on their own research, while the more applied fields of agricultural science were more likely to identify external factors influencing their field and problem choice.

In addition, it is often difficult to make a clear distinction between internal and external influences. On a general level all science is embedded in a specific cultural, social, and political milieu. As Husserl (1970:382) has suggested, "every science presupposes the life-world" (i.e., the world of everyday life). At another level, societal concerns (e.g., productivity, efficiency) may become embedded within the sciences to the degree that they become disciplinary standards by which research is judged. On a more specific level, while new instruments are developed outside a given science, or even outside science altogether (Ravetz, 1971:88-94), they are generally associated with methodological approaches and justifications that are fundamentally disciplinary in nature.

Additionally, the nature of the research methodology of the social sciences may lead to certain conclusions and hypotheses regarding the relative influence of internal and external factors. Questions regarding who had been most influential in choosing current research problems, how a subject had begun certain lines of research, how one achieved success in the field, and what influenced choice of research orientation were all likely to elicit internalist responses. In contrast, external explanations dominated rationale for the current hot topics and the avoided or ignored topics in each discipline, perceived audience for the research, and perceived considerations important for colleagues' research choices.

This suggests that internal and external influences are best conceptualized as different "strata" in the life-worlds of *both* the practicing scientist and the observer. Following Schutz and Luckmann it may be argued that our understanding of the world is divided into "finite provinces of meaning" (1973:24).

The practicing scientist discussing his own work is immersed in the technical details and complexities usually referred to as "internal" influences. When asked about his colleagues, however, he is forced to "step back" and put them in a broader, "external" context. Similarly, the social scientist concerned with the specifics of the research process will tend to see the scientist as a free agent acting according to the dictates of scientific discipline; the social scientist concerned with sociohistorical trends will tend to see the scientist as a pawn in the chess game of history. Lamentably, very little work to date has attempted to meld these provinces of meaning together.

A broader epistemological issue arises from our use of interview techniques. Our informants were chosen, not for their status as spokesmen, but in order to maximize their heterogeneity. However, all interviewing potentially suffers from the fact that scientists may well be unaware of the influences on their work. This, of course, is the classic problem of false consciousness. Nevertheless, we submit that, as a group, most if not all of the important influences are likely to be identified, even though no single individual is aware of them all. Indeed, the same is true for the authors of this very paper. Surely there are blind spots in our perceptions; nevertheless, we must assert that as a whole social scientists can make truth claims. To deny this is to deny the possibility of any science whatever.

While science differs from other institutions in its commitment to "public knowledge" (Ziman, 1968), it nevertheless is susceptible—like other institutions—to direction and misdirection by the larger social world. Future studies of science must attempt, by means of concrete examples, to understand the complex way in which these multiple factors converge to construct that extraordinary social product we call science.

Finally, the evidence analyzed above clearly indicates the socially constructed character of agricultural science. It strongly implies, therefore, that research on the diffusion of agricultural innovations must address social and economic factors shaping the development of those innovations. In so doing the diffusion literature should examine how and to what degree both the innovations and the decisions to adopt them are embedded in particular socioeconomic contexts.

REFERENCES

Arnold, David O. 1975. Dimensional Sampling: An Approach for Studying a Small Number of Cases. *American Sociologist* 5:147-50.

Berry, Wendell. 1977. *The Unsettling of America.* San Francisco: Sierra Club Books.

Bohr, Niels. 1958. *Atomic Physics and Human Knowledge.* New York: John Wiley & Sons.

Busch, Lawrence. 1980. Structure and Negotiation in the Agricultural Sciences. *Rural Sociology* 45:26-48.

Carson, Rachel. 1962. *Silent Spring.* Boston: Houghton Mifflin Company.

Cheney, H.B. 1974. Roles of the Agronomist. *Agronomy Journal* 66:1-4.

Cole, Stephen. 1979. Age and Scientific Performance. *American Journal of Sociology* 84:958-77.

DeSolla Price, Derek. 1963. *Little Science, Big Science.* New York: Columbia University Press.

Dunlap, T.R. 1978. Science as a Guide in Regulating Technology: The Case of DDT in the United States. *Social Studies of Science* 8:265-86.

Evenson, Robert E., and Yoav Kislev. 1975. *Agricultural Research and Productivity.* New Haven, Conn.: Yale University Press.

Feyerabend, Paul. 1975. *Against Method.* London: New Left Books.

Fisch, R. 1977. Psychology of Science. In *Science, Technology and Society,* ed. I. Spiegel-Rosing and Derek DeSolla Price, pp. 277-318. London: Sage.

Forbes, E.G., ed. 1978. *Human Implications of Scientific Advance.* Proceedings of the XVth International Congress of the History of Science. Edinburgh: Edinburgh University Press.

Fujimoto, Isao, and Emmett Fiske. 1975. What Research Gets Done at a Land Grant College: Internal Factors at Work. Davis: University of California, Department of Applied Behavioral Science, Mimeo.

Fujimoto, Isao, and William Kopper. 1975. Outside Influences on What Research Gets Done at a Land Grant School: Impact of Marketing Orders. Paper presented at the Rural Sociological Society Meetings, San Francisco.

Gadamer, Hans Georg. 1976. *Truth and Method.* New York: Seabury Press.

Gaston, Jerry. 1978. *The Reward System in British and American Science.* New York: John Wiley & Sons.

Gilpatrick, John D. 1976. The Crisis in Chemical Control. *Phytopathology News* 10 (July):2-3.

Glaser, Barney, and Anselm Strauss. 1967. *The Discovery of Grounded Theory.* Chicago: Aldine Publishing Company.

Haberer, Joseph. 1969. *Politics and the Community of Science.* New York: Van Nostrand Reinhold.

Habermas, Jurgen. 1970. *Towards a Rational Society*. Boston: Beacon Press.
Hightower, Jim. 1973. *Hard Tomatoes, Hard Times*. Cambridge, Mass.: Schenckman Publishing Co.
Husserl, Edmund. 1970. *The Crisis of European Sciences and Transcendental Phenomenology*. Evanston Ill.: Northwestern University Press.
Knorr, Karin D., and Dietrich W. Knorr. 1978. From Scenes to Scripts: On the Relationship between Laboratory Research and Published Papers in Science. Paper presented at the annual meetings of the American Sociological Association, San Francisco.
Krige, J. 1978. Popper's Epistemology and the Autonomy of Science. *Social Studies of Science* 8:287-308.
Krohn, Roger G. 1977. Scientific Ideology and Scientific Process: The Natural History of a Conceptual Shift. In *The Social Production of Scientific Knowledge*, ed. Everett Mendelsohn, Peter Weingart, and Richard Whitley, pp. 69-99. Boston: D. Reidel Publishing Co.
Kuhn, Thomas. 1970. *The Structure of Scientific Revolutions*. 2d ed. Chicago: University of Chicago Press.
_____ . 1977. *The Essential Tension. Selected Studies in Scientific Tradition and Change*. Chicago: University of Chicago Press.
Lewontin, Richard. 1976. The Problems of Lysenkoism. In *The Radicalization of Science*, ed. Hilary Rose and Steven Rose, pp. 32-64. New York: Holmes & Meier Publishers.
Luria, S.E. 1973. On Research Styles and Allied Matters. *Daedalus* 102:75-84.
Maritain, Jacques. 1944. *The Dream of Descartes*. New York: Philosophical Library.
Merton, Robert K. 1967. *On Theoretical Sociology*. New York: Free Press.
_____ . 1973. *The Sociology of Science*. Chicago: University of Chicago Press.
Minckler, Leon S. 1976. Directions of Forest Research in America. *Journal of Forestry* 74:212-16.
Nagi, Saad, and Ronald G. Corwin. 1972. The Research Enterprise: An Overview. In *The Social Contexts of Research*, ed. Saad Nagi and Ronald G. Corwin, pp. 1-27. New York: John Wiley & Sons.
Nicholson, Heather Johnston. 1977. Autonomy and Accountability in Basic Research. *Minerva* 15:32-61.
Picou, J. Stephen; Richard H. Wells; and Kenneth L. Nyberg. 1978. Paradigms, Theories, and Methods in Contemporary Rural Sociology. *Rural Sociology* 43:559-83.
Radnitzky, Gerard. 1973. *Contemporary Schools of Metascience*. 3d ed. Chicago: Henry Regnery.
Ravetz, Jerome R. 1971. *Scientific Knowledge and Its Social Problems*. New York: Oxford University Press.
Rose, Hilary, and Steven Rose. 1976. The Problematic Inheritance: Marx and Engels on the Natural Sciences. In *The Political Economy of Science*, ed. Hilary Rose and Steven Rose, pp. 1-13. New York: Holmes & Meier Publishers.
Rosenberg, Charles E. 1976. *No Other Gods: Science and American Social Thought*. Baltimore, Md.: Johns Hopkins University Press.
_____ . 1977. Rationalization and Reality in the Shaping of American Agricultural Research, 1875-1914. *Social Studies of Science* 7:401-22.
Saint, William S., and E. Walter Coward, Jr. 1977. Agriculture and Behavioral Science: Emerging Orientations. *Science* 197:733-37.
Schutz, Alfred, and Thomas Luckmann. 1973. *Structures of the Life-World*. Evanston, Ill.: Northwestern University Press.
Whitley, Richard. 1974. *Social Processes of Scientific Development*. London: Routledge & Kegan Paul.

Ziman, John M. 1968. *Public Knowledge.* Cambridge: Cambridge University Press.
Zuckerman, Harriet A. 1978. Theory Choice and Problem Choice in Science. *Sociological Inquiry* 48:65-95.

Part III

The Exportation
of Development

6

The Agricultural Sciences and the Modern World System

Lawrence Busch and Carolyn Sachs

And we make (by art) in the same orchards and gardens, trees and flowers to come earlier or later than their seasons; and to come up and bear more speedily than by their natural course they do. We make them also by art greater much than their nature; and their fruit sweeter and greater and of differing taste, smell, colour, and figure, from their nature.

Sir Francis Bacon, The New Atlantis, 1627

The age of the Utopia of Sir Thomas More and of the New Atlantis of Bacon, divested of fantasy and clothed in the habiliments of decorous sobriety, seemed to have dawned upon mankind.

J.K. Patterson, in a presidential address to the Association of American
Agricultural Colleges and Experiment Stations, 1903

In his utopian novel *The New Atlantis* Sir Francis Bacon (1974) described a marvelous world in which the members of the House of Salomon, in collaboration with the state, insured the continued agricultural prosperity of the larger society. In the 350 years that have passed since the publication of Bacon's work, the general outlines of his ideal society have been realized. The corporate body of agricultural scientists that Bacon described has become a nearly universal feature of the modern world system.

In this chapter we first explore the development of the agricultural sciences in the late nineteenth century. In so doing, we show how the agricultural sciences were initially linked to the spread of the modern world system and how they played a major role in the transformation of farming from a subsistence to a capitalist enterprise. Then, we show how social-science research was added to the system as a mode of improving research and extension systems while still taking certain basic assumptions about science and agriculture for granted. The third part of the chapter focuses upon the "new" international research centers and the essentially unpublicized ideological splits within them. Finally, the concluding section points to some of the more promising developments that might serve gradually to open the system to questions.

THE RISE OF AGRICULTURAL RESEARCH

As Wallerstein (1972, 1974) has noted, agriculture played a primary role in the creation of what he has termed the "modern world system." Starting about 1450 there have emerged three types of states: the core, the semiperiphery, and the periphery. The core consists of those states in which both labor and capital are most highly remunerated and where the most capital-intensive production takes place. The semiperiphery consists of those "older" states no longer part of the core. "The periphery . . . is that geographical sector . . . wherein production is primarily of lower-ranking goods (that is, goods whose labor is less well rewarded) but which is an integral part of the overall system of the division of labor, because the commodities involved are essential for daily use" (1974:301-2). Moreover, peripheral areas tend to be monocultural; in other words, each peripheral state tends to produce one or two cash crops for the benefit of the world economy (de Janvry, 1975; Chilcote, 1974).

From the 17th century to the present, more and more of the world has been incorporated into the modern world system. Farming has been fundamentally changed from a means of subsistence to a capitalist enterprise. As a result the volume of agricultural commodities grown for export has increased steadily. Until the late 19th century, however, much of the increase was attained through the introduction of new crops and by increasing the land area devoted to the production of export crops.

Brockway has noted the important role that science played even in this. She estimates that by 1800 the various European states supported over 1600 botanic gardens (1979:74). She further suggests that by 1800 "every new plant was being scrutinized for its use as food, fiber, timber, dye, or medicine. Botanic gardens consciously served the state as well as science, and shared the mercantilist and nationalist spirit of the times" (1979:74-75). French botanists helped introduce grapes and citrus to Algeria. British botanists from Kew Gardens illegally smuggled cinchona plants and seeds out of Bolivia, Ecuador, and Peru and used them to establish large plantations in India. The quinine produced in this way was of significant aid in maintaining British control over malaria-infested colonies. Similarly they removed rubber *(hevea)* seeds from Brazil and used them to establish plantations in Malaya, Ceylon, and India. Thus, "the colonial gardens functioned as agricultural experiment stations, doing the developmental work, educating the planters by personal contact and published bulletins" (Brockway, 1979:165).

Indeed, the botanic gardens were creatures of the state. There was "no way to draw the line between science, commerce, and imperialism in the work of Kew collectors" (Brockway, 1979:84). Yet, on the other hand, the early work of these gardens focused upon testing varieties in new climes and developing new methods of cultivation.

In the United States, this work was carried on by a series of government-sponsored expeditions, each of which returned with a wide variety of exotic species and varieties. "New varieties of cotton seed, corn, wheat, and other grains and vegetables, whether imported or developed in the United States, were given much attention by the farm journals and, if successful, were quickly adopted" (Gates, 1960:301). In short, much of the work of the botanic gardens (and the Patent Office in the United States) was accomplished by simple trial and error.

It was not until the advent of Liebig's agricultural chemistry that major increases in yield per hectare became thinkable. His linking of minerals in the soil to plant growth and his later work on fertilizers appeared to open a whole new world. During the 1850s his work was a constant topic of discussion in agricultural journals and lent enormous credence to the idea that science could remake agriculture.

Nevertheless, the nonproprietary character of improved seeds and methods of cultivation limited private support for research that would increase productivity per hectare. As Evenson and Kislev put it, referring to sugar research, "it soon became clear that it is not profitable to make large investments in private effort because the plantation was unable to capture more than a small fraction of the benefits" (1975:48). However, the competitive character of export markets encouraged decreased costs and increased yields. The worldwide experiment-station movement of the late 19th century provided a solution to the dilemma by making research the province of the state, or, somewhat less desirably, of a growers' association.

Through the establishment of experiment stations, growers hoped yields could be increased regionally and profits increased at the expense of other regions, while the costs of developing innovations could often be passed on to the state. Moreover, once one region successfully implemented an innovation, growers in other areas had to follow suit in order merely to hold on to their share of the market. Detailed worldwide comparative information regarding the origins, foci, and size of experiment stations is unavailable. However, some quantitative data for the world as a whole is available in various directories.

Starting in the mid-1850s, experiment stations were rapidly developed in Germany. As Table 6.1 reveals, France soon followed the German lead. The British were somewhat slower, preferring instead merely to expand the function of some of the botanic gardens that were already operational. Then, during the last quarter of the century, the United States, Russia, and Japan followed suit. By 1900 over 800 stations of varying size and competence were in operation around the world (True and Crosby, 1904; USDA, 1903).

During the period from 1900 to about 1930, the number of experiment stations increased to over 1,400 worldwide (Table 6.2). Nearly a third of these

---

Table 6.1   The Establishment of Experiment Stations, 1850-1900

	World total	United States	Total active stations British Empire	French Empire	Germany	Russia	Japan
Pre-1851	18		16				
1851-1855	24		16	1	4		
1856-1860	36		18	1	13		
1861-1865	49		20	2	19	1	
1866-1870	70		20	7	28	2	1
1871-1875	125	2	23	13	45	2	1
1876-1880	170	6	28	21	53	5	1
1881-1885	250	19	39	34	56	8	13
1886-1890	390	46	57	57	61	16	36
1891-1896	499	49	72	80	68	48	48
1896-1900	591	51	94	84	75	72	55

Note: Figures include only those stations for which founding dates were available. Thus, the growth of the movement is somewhat understated.
Source: A.C. True and D.J. Crosby, *Agricultural Experiment Stations in Foreign Countries,* Office of Experiment Stations Bulletin 112, revised (Washington: U.S. Department of Agriculture, 1904); USDA, *Annual Report of the Office of Experiment Stations* for the year ended June 30, 1902 (Washington: U.S. Government Printing Office, 1903).

---

were to be found in Europe and another third in Asia. Moreover, by 1930 nearly all British colonies and most French colonies had at least one station.

In general, experiment stations in core countries (or countries settled by migrants from the core, e.g., Australia) tended to be oriented toward food-crop production, while those in the periphery tended to focus upon export crops. Of course, these are overlapping categories: a food crop may be an export crop as well. However, it appears that this conceptual overlap only found concrete manifestation in the core and semiperiphery. In the periphery, food crops were rarely exported. Instead, "luxury" goods such as sugar, tea, and coffee were the objects of both research and export. In the data reported in Table 6.2, we have categorized all food under that heading regardless of whether it was grown for export or for local consumption. That table also

Table 6.2 Active Experiment Stations ca. 1930 by Location in the World System and Research Focus (N = 1422)

Location		Research focus			
Place	Number	Example	Total	Food	Export
Europe	479				
Britain	30	Britain	30	25	1
France	40				
Other core	118	Sweden	5	4	0
Soviet Union	54				
Semiperiphery	122	Hungary	16	7	5
Periphery	115	Portugal	12	3	8
Africa	225				
British Empire	73	Nigeria	12	3	6
French Empire	70	Ivory Coast	11	0	9
Other	82	Belgian Congo	18	0	14
		South Africa	23	11	10
Asia and Oceania	476				
British Empire	276	Malaya	24	7	9
French Empire	5				
Japan	54				
Other	141	Dutch Indies	8	0	7
Americas	242				
British Empire	71	Trinidad	7	1	3
United States	85[a]				
Others	86	Peru	17	10	4

[a] Includes only state experiment stations.

Source: International Institute of Agriculture, *Les Institutions d'experimenta-tion agricole . . .* 1933, 1934, *Les Institutions de laiterie . . .* 1934 (Rome: IIA); United States Department of Agriculture, *Workers in Subjects Pertaining to Agriculture in State Agricultural Colleges and Experiment Stations,* Misc. publ. 100 (Washington: U.S. Government Printing Office, 1931).

provides several examples, chosen for data availability.* For instance, in Great Britain, a core country, 25 of the 30 stations reported a focus on food crops as compared to one station that conducted research on tea, coffee, and rubber. In a semiperipheral country, Bulgaria, the ratio of food-crop to export-crop experiment stations was 3:1. Data from Portugal reveal the concentration of research on export crops in the periphery even in Europe.

Outside of Europe, agricultural research in the 1930s was directed primarily toward export crops. Especially in the British, French, and Belgian colonies of Africa, it is apparent that little emphasis was placed on research on local food crops. Indigenous populations that might have been concerned with food crops had little or no political clout, while wealthy capitalist farmers growing for export were able to apply pressure for support for research. For example, Evenson and Kislev note that sugarcane experiment stations were generally established in countries where grower organizations were strong (1975:48). Similarly, Ayer and Schuh (1972) note that cotton growers in the state of Sao Paulo, Brazil, through their control of the legislature, were able to funnel more state funds to cotton research during the 1930s than were devoted to the entire American hybrid corn program of the same period! A similar emphasis on export crops was found throughout the periphery. The exceptions seem to be those countries, such as South Africa and Peru, that were "independent." There, some experiment stations were oriented toward export crops, but the majority were devoted to food-crop research.

Available descriptive material supports the quantitative overview provided in Tables 6.1 and 6.2. For example, the bias in favor of the owner of the large capitalist farm or plantation oriented toward export markets was apparent at the beginning of the American experiment-station movement. Congressman Hatch was to argue that experiment stations were needed in order to insure the U.S. lead in agricultural exports (Hatch, 1886:2). Moreover, once established, "the political needs of the station scientists guaranteed that the educated, adequately capitalized farmer would be their natural ally in the achievement of power. Indeed, the larger the scale of the enterprise, the more likely it was— in general—to find experiment station scientists relevant. Innovation and adoption implied both capital and appropriate values" (Rosenberg, 1971:18).

Early American agricultural research tended to focus on grains. This was in large part the result of an historical and climatological accident that made food grains the major American export crops (Friedmann, 1978). By contrast, in British colonies, central research institutes were established focusing on specific export commodities such as cotton, coffee, or cocoa (Moseman, 1970:57). Indeed, the British did become interested in increasing Indian wheat production, but only after the opening of the Suez canal made it appear that it might be possible to undercut American wheat prices (Spitz, 1975:6). In

_____

*Available information is sketchy. The examples provided appear typical of what data are available, however.

Australia food-grain research was supported (Black, 1976). There, however, as in the United States, there was a large land mass suitable to grain cultivation and no large indigenous labor force ready to be exploited. Importantly, the possibility of importing Asian laborers to work on plantations in the tropical part of the country was considered; it was rejected, as it was thought that this would overly disturb the existing (European) settlement patterns (see South Australia, 1887:35-36).

Elsewhere within the British empire, it was not until the 1940s that work on food crops was seriously contemplated as a result of belated recognition of nutritional problems. "It was considerations of this kind that led the Colonial Agricultural Service . . . to step up research on food crops as compared with that on cash crops which had so preoccupied the staff at an earlier period" (Masefield, 1972:70). Moreover, this shift in interest was made possible by the weakened link between the planters and the state and the consequent drift from state to planter support for commodity research beginning in the 1920s (Masefield, 1972:78).

Agricultural research in French colonies was also directed toward metropolitan industrial markets. It was conducted on a smaller scale than that of the British due to the far more limited funds that the French had at their disposal. Even research on peanuts in Senegal was motivated by a concern for the French soap industry. "As for food crops, the French approach was generally to tell the chiefs to ask their people to plant more millet" (Klein, 1979:84-85).

While the Belgians never developed a large empire, Leopold saw science as the key to European conquest of Africa. In 1879, Stanley led an expedition up the Congo River for Belgium. One aim of that expedition was the establishment of gardens to see what crops could be profitably grown. By the end of the first decade of this century the Belgians had appointed agronomist Edmond LePlae as full-time Director-General for Agriculture. LePlae encouraged close collaboration between the state and private enterprise. Moreover, he encouraged the forced growing of cotton: "In countries with very backward agriculture, the temporary requirement of obligatory plantings is often necessary to assure the indigenous population sufficient and regular nourishment and to introduce the export crops that will be the principal sources of prosperity and well-being for the natives" (quoted in Cornet, 1965:138; our translation). By 1939, 93 Europeans and 6670 Africans were employed by the *Institut Nationale pour L'Etude Agronomique du Congo Belge,* working on the improvement of a wide range of export crops (Cornet, 1965:189). Moreover, even this export-oriented structure collapsed at independence with the mass exodus of trained personnel.

A somewhat different situation was to be found in Dutch-controlled Java. There, population density limited the production of cash crops. Moreover, Java was a net food importer. As a result, experimental work was divided into "crops for native culture and those for European culture" (Russell, 1927:116; see also Palmer, 1972b). The former were grown as food crops, while the latter were grown for their commercial value. Moreover, a government-financed station devoted itself to food crops—perhaps to insure that population growth could be accommodated without expanding the land area utilized for food production—while planters financed crop-specific private experiment stations. The sugar station was, in the mid-1920s, the largest of its kind in the world, with a budget of $500,000 per annum, all derived from private sources (Russell, 1927:117).

There were, of course, those Third World countries in which food-crop research designed to be beneficial to small farmers was undertaken. However, all too often this was only a short-lived experience. Mexico is a case in point. During the 1930s an Office of Experiment Stations was established by the Cardenas government and was intimately related to the land reform. Scientists at this institution "were little interested in importing technology from abroad, preferring to work slowly at the local level with land reform beneficiaries in an effort to find solutions to the practical problems faced daily by the latter" (Hewitt de Alcantara, 1976:19). Their work, however, was gradually reduced in importance after 1941 when the Mexican government, in conjunction with the Rockefeller Foundation, established a research program aimed at providing a surplus that could feed the cities rather than resolve the problems of subsistence farmers. In 1961, the two organizations merged and foreign scientists officially withdrew, "but the heritage of a sophisticated research effort for commercial agriculture continued to orient the work of the new national organization, and the long-standing lack of commitment to attacking the problems of near-subsistence agriculture through scientific investigation has only recently begun to change" (Hewitt de Alcantara, 1976:20).

Japanese policy toward their colonies also reflected this export emphasis: "The official economic policy of the Japanese administration in Taiwan emphasized expansion of sugar production rather than rice production during the first two decades of the colonial period" (Evenson, Houck, and Ruttan, 1970:462). In addition, U.S. policy toward its Latin American neighbors was oriented in this direction. Writing of the establishment of the first cooperatively run experiment station in the Americas, Moore observed that these and other stations were primarily "designed to promote profitable production of... export crops (1943:107). Nor had the situation appeared to change by the mid-1970s. Summing up the current situation in tropical areas—where most export crops are currently grown—one observer notes that "it is repeatedly stated that tropical staples are ignored in research programs, while export crops are studied extensively" (Janzen, 1975:107). One result of the bias toward research on export crops is that in the tropics "maize is now generally relegated to the poorer lands, because the better lands are frequently devoted to cash crops such as cotton, sugar cane, and coffee" (Wellhausen, 1975:61).

Recent worldwide evidence is more difficult to come by. Some information is summarized in Table 6.3. That table contrasts the number of crop-specific

Table 6.3   Crop-Specific Agricultural Serials in Print for Selected Crops

SITC Number	Crop	World market economies value exported—1974 ($1,000)	Number serials 1970	Number of nations w/serials (N = 66)
	Cereals			
041	Wheat	9,648,886	32	13
042	Rice	2,403,941	77	20
043	Barley	1,523,399	12	8
044	Maize	6,101,788	31	14
—	Sorghum/millet	N.A.	2	1
—	Oats	N.A.	5	2
	Export Crops			
0513	Bananas	684,041	7	6
061	Sugar	7,055,909	160	35
071	Coffee	4,674,580	77	24
072	Cocoa	2,252,667	29	10
074	Tea	799,223	52	14
121	Tobacco	2,208,886	113	33
2311	Rubber	3,140,349	33	11
263	Cotton	4,296,779	118	25

Source: D.H. Boalch, ed. Current Agricultural Serials (Oxford: International Association of Agricultural Librarians and Documentalists, 1965); International Association of Agricultural Librarians and Documentalists Quarterly Bulletin, various issues 1966-1970; United Nations, Yearbook of International Trade Statistics, 1975 (New York: United Nations, 1976).

journals published in 1970 for each of the major food grains with those focusing on a number of the major export crops. It is immediately apparent that as of a decade ago a great deal more effort was expended on export-crop research than on research relating to food grains. Moreover, the data reveal that, within most countries, research on export crops predates that on food grains. Indeed, in half the nations studied, no grain-related serials were published. Of course, it may be readily conceded that such indicators are quite crude and that it would be unwise to employ them as a means of gauging *national* research priorities. However, for the world as a whole the pattern is remarkably clear. It appears that, historically, the research goals of agricultural sciences have been strongly influenced by the needs of the modern world system and have been little concerned with either food production or the needs of subsistence farmers. As Susan George has put it, agricultural research "only got under way ... because settlers introducing cash crops into newly colonized areas found their plants being attacked by myriad diseases and pests in unfamiliar environments and the planters themselves were being wiped out financially. Research stations sprung up throughout the colonial world, but predictably paid no attention whatever to local food crops. This research lag between cash and food crops is, alas, still with us" (1977:67).

Does this not, however, merely imply that research expenditures in agriculture must be redirected toward food production? Moreover, doesn't it suggest that expenditures on agricultural extension need to be increased so that benefits of research will reach even the smallest farmer? Put another way, are we really talking about a problem of science or are we instead concerned with the social system that provides its funds and diffuses its results?

Such questions cannot be answered at the global level. They require an inquiry into the sociopolitical context of specific development schemes. While it is impossible to present more than a sketch here, a look at one such project should serve to clearly raise the issues, even if it cannot hope to fully resolve them.

The Gezira Development scheme in the Sudan provides us with what is probably the most well-documented and long-standing British colonial agricultural "development" scheme currently functioning. It is particularly relevant as it was to be a development scheme and not merely an exploitative reorganization of agricultural production such as those in 18th-century India. The Gezira refers to an area of land located between the Blue and the White Nile. This hot, dry, and generally flat plain has been under irrigation for some 50-odd years. While it has been considered highly successful by many, it has recently come under considerable attack. According to Barnett, "the major factors in the establishment of the Gezira scheme were not only the decline of the British cotton industry but also the requirements of the imperial grand strategy" (1977:4). In brief, it appears that the scheme was established in large part in response to the owners of the Lancashire cotton mills. They needed a steady supply of high-quality cotton not available in either the Egyptian or Indian colonies. The Gezira, when properly irrigated and planted with improved varieties, promised a steady supply of cheap, high-quality cotton. This scheme, like virtually all other smallholder irrigation projects, has forced a highly authoritarian organizational structure upon the residents. In addition, it has made the nuclear family the relevant unit of production and forced a form of possessive individualism upon the tenants (Barnett, 1977:89, 96). Moreover, at the national level, "this kind of dependence upon cotton monoculture places the Sudan in a precarious economic position in terms of its trading (not to say political) relations with other countries" (Barnett, 1977:14).

In considering the example of the Gezira, we might also ask: What role did the agricultural sciences play in its development? It is curious that we must go to an early, and virtually acritical, work by Gaitskell (1959) to find an answer to our question. As that author put it:

In 1918 a Gezira Research Farm had been started near Wad Medani, the capitol of the Blue Nile Province, to study soil and water management, crop varieties, rotations, cultivation practices, fertilizer response and, of course, diseases and pests. With the setting up of this scientific station in the heart of the Gezira there began a close association between the backroom boys of the research farm and the field staff of the Syndicate, not at all times easy but always stimulating and destined to play a vital part in the survival of the scheme (1959:138).

This research farm, financed by the government with a contribution from the privately owned Syndicate, was connected to the "Empire Cotton Growing Corporation" as well as to the Rothamsted Experimental Station in Britain (Gaitskell, 1959:139). Indeed, a former officer in the Colonial Agricultural Service has argued that "the Empire Cotton Growing Corporation had links so close that many of its staff carried out their own work at the experimental stations of colonial Departments" (Masefield, 1972:7). In short, the agricultural research performed in the Gezira, without which the entire scheme would have been unworkable, can in no way be regarded as the work of disinterested practitioners. Instead, it must be regarded as fully integrated into and supportive of the modern world system. The researchers' link to the world system is further revealed in that, when the country was turned over to the Sudanese, most of the researchers left (Gaitskell, 1959:238).

Nor can we regard the Gezira experience as aberrational. The Volta River project in Ghana provides us with yet another illustration. There we find that "in addition to choosing suitable soils, using improved seeds, and applying fertilizers, manures, insecticides and fungicides, the *agriculturalists* were convinced that the way to obtain the required high increases in field crop production was to introduce mechanization" (Kalitsi, 1970:42; emphasis added). Indeed, they initially assumed that with resettlement, farmers would switch from their complex intercropping systems to the growing of a single, mechanized cash crop (Chambers, 1970:236). One observer has noted that "one is entitled to wonder what this degree of mechanization would leave for the farmer to do" (Kalitsi, 1970:42). While the technical problems proved virtually insurmountable, it is clear that scientific work was directed toward the substitution of a single export crop for the balanced crop production typical in most African villages.

Similarly, both Belshaw (1969:18) and Moris, in discussions of recent resettlement schemes in East Africa, have noted that agricultural research there has emphasized cash crops, capital-intensive methods, and mechanization. Finally, Sorrenson (1968), reflecting on the origins of European settlement in Kenya, has noted a similar pattern in agricultural research conducted there.

In short, while the evidence is necessarily sketchy, it appears that agricultural research has tended to favor export crops over food crops and capital-intensive over labor-intensive methods. This, in turn, has tended to (1) increase personal and national dependency upon the vagaries of the world commodity markets, and (2) concentrate power and wealth in the hands of those who can afford the high-cost inputs. Moreover, large numbers of individuals have been forced off the land, contributing to already high urban unemployment.

In recent years, however, particularly since the end of the second world war, the scope—if not the goals—of agricultural research has changed significantly in at least two ways: social scientists have made their presence felt, and international research centers focusing upon food crops have been developed. It is to these new developments that we now turn.

THE ENTRY OF SOCIAL SCIENCE INTO AGRICULTURAL RESEARCH

During the nineteenth century, agricultural research was almost wholly confined to the biological and physical sciences. However, by the end of the second decade of this century, social and economic research had become a noticeable feature of the American experiment stations. Despite this, social scientists with few exceptions tended to play a rather narrowly delimited role. Specifically, they were charged with (1) helping farmers become more efficient by applying the principles of farm management, (2) developing marketing information and forecasts for a wide range of commodities, and (3) helping to encourage the broad diffusion of agricultural innovations.

These three avenues of research were linked by an essentially acritical enthusiasm for the products of the technical sciences. Many sociologists and economists in land-grant institutions shared with their colleagues in more technical fields a belief in the superiority of scientific knowledge. Indeed, a not insignificant number of agricultural economists and rural sociologists were initially drawn from the more technical disciplines. As such, they tended to see their role as limited to helping farmers adjust to a changing agricultural scene.

If farmers could not individually raise the prices they received for their products, improved farm management would help lower production costs. Moreover, better marketing information would, it was argued, make the agricultural sector operate in practice in a way more congruent with economic theory. If perfect information was an unattainable ideal, at least substantial improvements in the availability of information could be made. These improvements, it was felt, would help farmers respond to market fluctuations.

A third research trajectory arose with respect to the diffusion of innovations. Its proponents were concerned with improving the flow of communications between scientists and farmers so as to speed up the adoption and diffusion of agricultural innovations (e.g., Rogers and Shoemaker, 1971).

Moreover, like their colleagues in marketing and management, many researchers concerned with diffusion tended to see tradition as diametrically opposed to modernity and social structure as an essentially unchanging context that limited the goals and aspirations of isolated individuals (cf. Busch, 1978).

In contrast to those who pursued the three research traditions described above, a small but dynamic group of social scientists did begin to address issues of social and economic policy at about the same time. However, they quickly incurred the wrath of certain powerful interests, particularly the Farm Bureau (Friedland, 1979; Kirkendall, 1966). By the early 1940s, most of those concerned with agricultural policy issues had either resigned under pressure or moved to less controversial areas of research. Only a small group outside the public sector, and concentrated at the University of Chicago, remained (Hardin, 1955).

Thus, as the world emerged from World War II, there were few social scientists in the United States with an interest in matters of agricultural policy. At the same time, American foreign-policy makers were holding up the American model as the solution to the problems of underdevelopment. The Marshall Plan appeared to be the appropriate catalyst to create the conditions necessary for Rostow's (1960) metaphorical economic "take-off." Yet, as Chirot has argued, "If anything, American aid of the 1945-1955 period was too successful. It fully accomplished its aim of preserving the world capitalist core from communism and from a relapse into the aggressive nationalist excesses of the 1930s . . . . An illusion was created that massive aid could work under different circumstances, outside the core as well as in it" (1977:150). Of course, agricultural aid was effective in Europe for the same reasons that aid in general was effective: European countries had well-educated, highly skilled populations and were true nation-states. At the time no one seemed to notice that the "nations" of the periphery lacked these essential features. Thus, aid tended to stress rapid industrialization through capital investment.

When it became apparent that merely providing agricultural capital was insufficient for developing a "modern" agriculture, U.S. policymakers turned to agricultural social scientists for a solution. During the 1950s and 1960s social scientists became increasingly involved in attempts to diffuse to the Third World agricultural innovations developed in the West. At first, "rather than examine whether the new technology was indeed appropriate and productive, social scientists and practitioners all too often tried to explain why people in LDC's did not adopt it" (Uphoff, Cohen, and Goldsmith, 1979:285-86). They noted again and again the tendency for innovations to be adopted more often by larger, higher-status, better-capitalized farmers. Only somewhat later did they begin to reflect on the nature of the innovations themselves. Perhaps, some suggested, the innovations were not well-adapted to the situation found in less-developed countries. From this concern arose a new approach in the late 1960s: institution building.

There were several central themes that together constituted institution building. First, it was increasingly argued that the products of Western agricultural research could not be directly transferred to the Third World. Instead, it was necessary to perform "adaptive" research in each country so as to make research results relevant. Second, it was necessary to establish a network of formal organizations in each country capable of providing agricultural research, extension, and education. Those organizations that did exist, it was asserted, were usually poorly organized and ineffective. The land-grant model that appeared so intimately related to increased productivity in American agriculture would be reproduced in developing countries.*

Economists were particularly active in developing the political support necessary for an institution-building perspective. Econometric studies of (1) the impact of research and extension on productivity, (2) the effectiveness of research resource allocation in increasing productivity, and (3) the economic factors creating "induced innovations" in agriculture rapidly multiplied (e.g., Fishel, 1971; Arndt, Dalrymple, and Ruttan, 1977; Binswanger and Ruttan, 1977).

Initially, the institution-building model was unabashedly elitist in its perspective. It was assumed that peasants were bound to outmoded traditions and could only be pried loose by a massive organizational effort (Cochrane, 1972). Moreover, as one of its leading proponents explained it, "The IB model is an elitist theory with an explicit social engineering bias. Changes occur from the top down, not from the bottom up, and they are guided by persons enjoying a measure of official authority or sanction" (Esman, 1972:66). Such a position, of course, assumed that elites were, if not democratically elected, at least honestly concerned with national development more than with personal aggrandizement. It also assumed that the social and natural sciences could provide all the answers relatively quickly and easily if the institutional conditions were right. Clearly, in many cases, both assumptions have been proved false.

EXCURSUS: THE ROLE OF FOOD AID

At the same time that an attempt was being made to diffuse agricultural innovations and build institutions in underdeveloped countries, food-aid policies were altering the productions systems in the less-developed countries. While this subject is somewhat peripheral to our main theme, it cannot be ignored, for food aid has often inadvertently resulted in less-developed nations becoming increasingly dependent on imported food. As large quantities of low-priced imported grain enter the markets of less-developed countries, local production is often reduced. While food aid is an effective

---

*This was certainly not the first time the land-grant model had been copied. As early as 1875, it was actively sought out by Japan (Butterfield, 1923). However, it was only actively promoted with the institution-building movement.

short-term answer to acute food shortages, in the long run, food aid leads to increased dependency (Dumont and Cohen, 1980).

In order to comprehend the precariousness of this dependency for the less-developed countries, it is necessary to remember that food aid is surplus food from the developed countries. Thus, the greater the surplus in developed countries, the lower the prices importers pay. If the developed countries should fail to produce large surpluses, the poorer countries will be the first to go without food. This could result in disaster in those countries which have reduced local food production due to the low prices of imported food. Moreover, the major supplier of food aid is the United States, which has frequently distributed food aid for political objectives. Nations seldom receive food with no strings attached. Therefore, this increased dependency does not bring stability to the poorer countries. An additional problem with food aid is that even if the food reaches a country, there is no guarantee that the hungriest people will receive the food (Dumont and Cohen, 1980). Poor transportation networks and corruption often prevent the transfer of food to the people who have the least.

THE DEVELOPMENT OF THE INTERNATIONAL AGRICULTURAL RESEARCH CENTERS

With attempts at development in Third World countries directed toward industrialization or agricultural goods for export, research on the staple crops of many countries was largely neglected. The pressures of urban population growth and the low level of food production in many countries brought the Malthusian issue to the forefront. In 1941, the Rockefeller Foundation sent a team of scientists to Mexico to experiment with increasing grain production. These scientists were primarily plant breeders who introduced high-yielding varieties (HYVs) of grain, which appeared to be immensely successful in increasing production. The implementation of this new type of agriculture making use of new seeds and dependent upon fertilizer, irrigation, and pesticides offered a promise that became widely known as the "Green Revolution." The ideology and research that formed the backbone of the Green Revolution also served as the model for the establishment of the International Agricultural Research Centers.

The first international research centers, the International Rice Research Institute (IRRI) in the Philippines and the International Center for the Improvement of Maize and Wheat (CIMMYT) in Mexico, were established in 1959 and 1964, respectively. IRRI was financed jointly by the Rockefeller and Ford foundations, while CIMMYT was established with $1 million from Rockefeller. These centers became the models for other international research centers. The focus of each was on several specific commodities, generally cereals. Dahlberg (1979) sees the development of the international research centers as the extension of the original Rockefeller initiative in Mexico in the early 1940s. As he states, "the research package in Mexico became a kind of blueprint successively applied to try to increase crop production, first on a geographic basis (national, regional, hemispheric, and globally) then on a climatic basis (within the tropics, then the semiarid tropics)" (1979:58).

In 1971, after four centers were successfully established, the Consultative Group on International Agricultural Research (CGIAR) was established. Currently, CGIAR administers 11 international research centers. In 1971, as at present, the World Bank provided the leadership for the group; other sponsors of CGIAR included the Food and Agriculture Organization (FAO) and the United Nations Development Program (UNDP). Membership included the Rockefeller, Ford, and Kellogg foundations, nine nations, two regional banks, and the International Development Research Center of Canada (Crawford, 1977). The primary objective of CGIAR was to encourage research to increase agricultural productivity in developing countries.

The international research centers were also to some extent modeled after the land-grant system in the United States. Both systems were predicated on the idea that agricultural research should be publicly funded. In order to transform traditional agricultural practices, the government in the United States established agricultural research centers in every state. The international centers have been established to transform agricultural production through science throughout the world. Both systems have tended to utilize scientific knowledge to transform agriculture into a more capital-intensive enterprise. The package of goods which has been introduced by the international centers is essentially the same as that which has been developed in the United States. One difference is that the international centers are more crop-specific than the U.S. agricultural science system. Schultz, who provided the philosophical justification for the international research centers, critiqued the U.S. agricultural-science establishment as being "plagued by too many substations and small federal installations" (1964:151). Thus, the international agricultural research system was set up to be more centralized than the American system. Moreover, the international centers receive the bulk of their funding and many staff members from the United States.

The Green Revolution, which was in large part a product of the international centers, has had many problems that have been well documented. While the plant breeders were successful in increasing the productivity of particular crops, they had little understanding of the differential impact of their high-technology package on the agricultural structure of various countries (Dahlberg, 1979). At the beginning, it was the countries that were not immediately faced with the Malthusian situation that were the centers for research. As Palmer (1972a) notes, the centers for wheat and rice breeding were located in Mexico and the Philippines—places where the natural environment was extremely favorable. Farmers with the best land, access to credit, and irrigation systems were the primary beneficiaries of the HYV technology package. Thus, the new products tended to favor the well-to-do farmers, while often increasing the impoverishment of the smaller farmers (Griffin, 1972; Frankel, 1971). In addition, farm size, land-tenure patterns, and distribution of wealth both affected and were affected by the adoption of HYVs. By generating the need for credit, Green Revolution technology actually increased the gap between rich and poor in many places. As this gap increased, rural impoverishment became more problematic than it had been prior to the introduction of HYVs. Rural unemployment increased; and people were often forced to move to urban areas, which resulted in urban population growth. The cities in developing countries had so few jobs available that many of the migrants were and remain unemployed or employed in low-paying jobs. In essence, changes in agricultural practices often contributed to both rural and urban poverty.

A member of the Technical Advisory Committee of CGIAR explained the early emphasis on production as follows: "It is fair to say that our concern was production, and we did not in our early work debate extensively the question of maximizing production versus optimizing farm income distribution, whatever that may mean" (Crawford, 1977:592). Nevertheless, he was somewhat hesitant to admit the importance of studying issues of distribution.

Recently, the problems associated with focusing on productivity at the expense of distributional issues have become so apparent that the international research centers have begun to incorporate equity issues into their goals. A recent report by the IRRI included a foreward by the director, who proclaimed the following goals for his center: "The IRRI's primary mission is to improve rice production and quality for the benefit of people in Asia and other rice producing countries. Our concern is not simply with rice yields, but also with their impact on the well-being of people both rural and urban and especially those with low incomes" (Brady in Hayami, 1978:xix).

As high-yielding varieties have spread throughout the world, single improved cultivars have tended to replace hundreds of traditional varieties. Moreover, plant breeders' attempts to improve yield have often been based upon improving one gene through cross-breeding (Palmer, 1972b). This emphasis on a single gene can weaken the composition of other genetic improvements that have been established over the centuries. Moreover, while the HYVs may produce more under optimal conditions, such as high levels of fertilization, adequate water, and protection from crop pests, they have tended (1) to have less stable *annual* yields, even though their *average* yields are higher, and (2) to perform poorly under less than optimal conditions. Recent crop failures, such as the southern corn leaf blight epidemic in the United States, attest to the problem of using seeds with limited genetic variability (Harlan, 1975). HYVs are often more vulnerable to pests as well (Wade, 1974).

As HYVs have replaced them, many traditional varieties have become extinct (Dahlberg, 1979), thereby diminishing the available gene pool. Yet, the preservation of genetic variability is a key to insuring that the long-term food needs of the world population are met. Fortunately, many of the international institutes consider part of their job to be the assemblage and preservation of germplasm for their particular crop(s). For example, CIMMYT houses a large collection of maize germplasm, while IRRI has expanded its rice collection. However, the actual success of these seed banks in preserving genetic diversity is questionable. As a recent article in *Science* states, "seed banks are in theory a reasonable way of halting genetic erosion, but in practice they tend to be underfunded, inadequate, and vulnerable to accidents or carelessness" (Wade, 1974:1187). In fact, some of the earlier and irreplaceable maize collections at CIMMYT were lost when the seed bank was reorganized in the 1960s. And, despite the creation of an International Board for Plant Genetic Resources, many scientists are concerned that the classificatory documents and computerized data banks do not accurately reflect the condition of the seeds themselves.

A further problem with the HYVs has been the focus on production at the expense of nutrition. While the International Research Centers focused their research on food crops, often the food needs of the poorer people went unmet. Palmer refers to the "final divorce of agriculture and nutrition through the agency of the 'Green Revolution'" (1972a:57). Plants were bred with primary emphasis on increased yield rather than on nutritional qualities. As a result, some of the new varieties actually had less protein content than the traditional varieties (Dahlberg, 1979). Moreover, the spread of HYVs has often reduced the variety of other foods available in local markets in many regions, having a detrimental effect on the nutritional adequacy of diets (Dumont and Cohen, 1980). However, as Ryan and Binswanger (1979) point out, the net nutritional impact of new varieties must be considered in addition to the protein content. In the semiarid tropics the major nutritional deficiencies are calories, vitamin A, vitamin B complex, and minerals. In these regions, the new varieties have resulted in improved nutrition in terms of protein and energy.

The problems connected with the Green Revolution have forced the international centers to focus on the issue of appropriate technology. The HYV package, which relies extensively on high-level technology and high-energy inputs, has proved itself largely inappropriate for small farmers in many developing countries. There is a limited amount of fertilizer and energy available, and the bulk of these products is consumed in Western countries. As farmers in developing countries utilize the HYV packages, they become increasingly dependent on inputs from the developed countries. It has become apparent that changes must be made so that technology is appropriate to local conditions.

THE CHALLENGE OF THE FUTURE

Some time ago, David Lowenthal (1960) vividly illustrated how three different cultures settled an area that is topographically and climatologically approximately the same. The contrasts between the Guianas serve to underscore the link between culture and agricultural science and technology. If we are to have a world that is meaningful, then we can no longer afford to accept the products of agricultural research as "undiluted goods." Nor can we afford to let efficiency remain the sole, or even the most important, measure of the value of research.

What factors, then, must an agricultural science policy include if it is to produce genuine development? Obviously, it must put food production ahead of the production of export commodities. The most recent available information does suggest some shift in worldwide research priorities. A comparison of the volume of scientific journal articles published annually from 1970 to 1978 for selected cereals (Figure 6.1) and export crops (Figure 6.2), though an admittedly crude measure, does show a trend toward more food-crop research with little change in the status of export-crop research. This is clearly a step in the right direction. However, as Heady (1971) and Friedland (1974) have suggested, at the very least the welfare implications of agricultural research must, insofar as is possible, also be made explicit before projects are undertaken. What this means is that a form of cost-benefit analysis must be developed that takes into account more than the returns to research in terms of increased productivity and includes factors such as farm family structure, the social role of particular crops, and the kind of social structure necessary to improve nutrition. Clearly, these latter factors may remain unmeasurable in economic, or even quantitative, terms.

Systems analysis, already in wide use within biology, presents one possible method for achieving this aim. However, it is well to remember that "As soon as we recognize that physical systems are embedded in, or interact with social systems, we recognize that science . . . can no longer be free from value judgements. Social systems involve not merely the interactions of physical forces but also contests of will arising from the purposiveness of behavior of animate elements in the system" (Dillon, 1976:7). In short, systems analysis cannot be used as a way of avoiding human judgement: it can only provide an aid in making such judgments. What is advocated here is not the creation of a "department of cost-benefit analysis" at every agricultural research institute. Such analyses will be of value only if they are fully integrated into the process of doing agricultural research. Put another way, agricultural research must be redefined so as to include far more than what goes on in the laboratory or experimental field. The subject matter of agricultural science must be reconstructed so that it includes social issues; the link between science and everyday life must be reestablished.

As Robert Seidel (1975) has suggested, the Third World suffers from the "burden of derivative modernization." The lack of research capabilities in the Third World forces it to be overly reliant on the sometimes inappropriate products of agricultural science produced in the West. Moreover, many Third World scientists, having received their training in the West, are socialized in

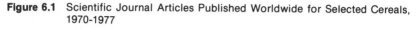

Figure 6.1   Scientific Journal Articles Published Worldwide for Selected Cereals, 1970-1977

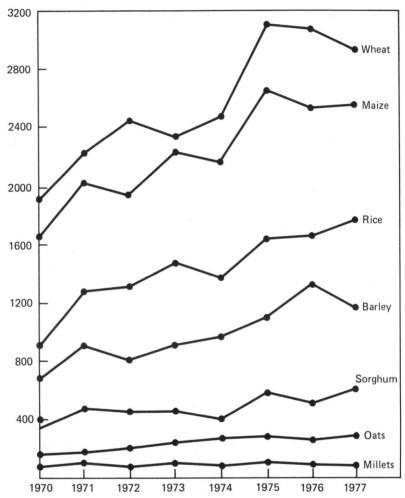

Note: Excludes those dealing with nutritional character of cereals.
Source: Agricola (Agricultural On-Line Access, National Agricultural Library).

such a way as to carry a great deal of Western intellectual "baggage" back to their homeland. As a result, what research is done in the Third World often suffers from the same inappropriateness characteristic of products borrowed from the West.

This brings us to that extremely popular topic of alternative or appropriate

Figure 6.2 Scientific Journal Articles Published Worldwide for Selected Export Crops, 1970-1977

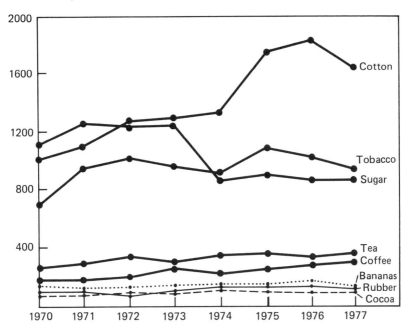

Note: Excludes those dealing with the nutritional character of crops.
Source: Agricola (Agricultural On-Line Access, National Agricultural Library).

technologies. Ever since the publication of E.F. Schumacher's *Small is Beautiful* (1973), there has been a surge of effort in the direction of so-called appropriate technology. Yet, as Dahlberg suggests in a recent paper (1978), we must not confuse alternative technologies with alternative systems. As Dahlberg notes, those organizations associated with alternative technologies tend to be occupied with short-term questions and are frequently unaware of the sociocultural implications of their technological solutions. By contrast, organizations with a systems orientation tend to have a longer time span for their research and to rely less on experiment-station trials. "Appropriate" technology is no more appropriate than any other technology if it avoids the sociocultural questions discussed above.

Some recent work at the international agricultural research centers appears to employ a systems approach, although even there it often meets with stubborn resistance (see, for example, Vallianatos, 1976). At least four of the centers are utilizing a farming systems approach, in which the interrelations between crops, animals, tools, soils, and workers are studied (CGIAR, 1978).

In this way, scientists are able to comprehend the small farmers' problems and needs, and their research can be oriented to benefit these systems.

Other scientists have begun to take seriously the dynamics of peasant agriculture. This is of importance in that, while traditional peasant agriculture may have low annual productivity, the risk is often low. The practices of the peasants have been established over centuries, and, although they may seem unscientific to Westerners, many of the practices are based on folk knowledge that is not immediately apparent to agricultural scientists. What this implies is that for agricultural research to produce both productivity and equity, farmers must become active participants in the process of doing science.* Moreover, they must be treated "not merely as the targets of advisory exhortation, as pupils at farmer training centers, or as the passive victims of development done to them by a remote government from afar: they have much to tell about soils, weather, crops, animals, diseases, and pests, as well as about their own purposes and difficulties" (Bunting, 1979:8). In short, a systems approach must involve an attempt to take farmers' categories and world views seriously rather than to impose our own upon them so that they can be fitted into our "system." Lest this point be misconstrued, it should be restated negatively. It would be absurdly romantic to suggest that agricultural research be abandoned or left to Third World farmers. It would also be naive to suggest a return to the bucolic world of some fantasized past. It is equally romantic, however, to see the products of agricultural research as a panacea that should be gratefully accepted by those same farmers. Cultural barriers do exist, but they can be overcome by dialogue (Busch, 1978) and a great deal of hard work.

One major attempt at research aimed at the development of alternative systems can be found in China. There, the development of a "mass scientific network" (Sigurdson, 1977:80) attempts to integrate research and extension in agriculture, public health, and industry. By centering improvement in technique and equipment at the village level, the Chinese appear to have kept costs down and benefits evenly distributed. One observer has noted that "local agricultural methods and varieties of seed have been upgraded to maximize output" (Erisman, 1975:340). On the other hand, it appears unfortunate the Chinese have devoted little attention to basic or long-term research (Sprague, 1975:58). While it is too early to assess the success of the Chinese experience, or even its desirability, this experiment does suggest that alternative approaches to agricultural science are within reach. If agricultural development is to reflect equity rather than inequity, the interests of society as a whole rather than those of the vested few, and meaningfulness and involvement rather than alienation, then a great deal more attention must be paid to the social construction of agricultural research.

*Currently, the United Nations Research Institute for Social Development (UNRISD) is inquiring into these issues as part of a global program on popular participation (see, e.g., Waddimba, 1979).

REFERENCES

Arndt, Thomas M.; Dana Dalrymple; and Vernon Ruttan, eds. 1977. *Resource Allocation and Productivity in National and International Agricultural Research.* Minneapolis: University of Minnesota Press.

Ayer, Harry W., and G. Edward Schuh. 1972. Social Rates of Return and Other Aspects of Agricultural Research: The Case of Cotton Research in Sao Paulo, Brazil. *American Journal of Agricultural Economics* 5:557-69.

Bacon, Francis. 1974. *The Advancement of Learning and the New Atlantis.* Oxford: Clarendon Press.

Barnett, Tony. 1977. *The Gezira Scheme.* London: Frank Cass.

Belshaw, D.G.R. 1969. An Outline of Resettlement Policy in Uganda, 1945-63. In *Land Settlement and Rural Development in Eastern Africa*, ed. R. Apthorpe, pp. 14-23. Kampala: Transition Books.

Binswanger, Hans P., and Vernon W. Ruttan. 1977. *Induced Innovation: Technology, Institutions, and Development.* Baltimore, Md.: Johns Hopkins University Press.

Black, Alan W. 1976. *Organizational Genesis and Development, A Study of Australian Agricultural Colleges.* St. Lucia: University of Queensland Press.

Boalch, D.H., ed. 1965. *Current Agricultural Serials.* Oxford: International Association of Agricultural Librarians and Documentalists.

Brockway, Lucille H. 1979. *Science and Colonial Expansion.* New York: Academic Press.

Bunting, A.H. 1979. *Science and Technology for Human Needs, Rural Development, and the Relief of Poverty.* New York: International Agricultural Development Service, Occasional Paper.

Busch, Lawrence. 1978. On Understanding Understanding: Two Views of Communication. *Rural Sociology* 43:450-74.

Butterfield, Kenyon. 1923. Agricultural Education and Research in China and Japan. Proceedings of the 36th Annual Convention of the Association of Land-Grant Colleges, Washington, D.C., pp. 56-60.

Chambers, Robert. 1970. Postscript and Discussion. In *The Volta Resettlement Experience*, ed. R. Chambers, pp. 226-69. New York: Praeger Publishers.

Chilcote, Ronald H. 1974. Dependency: A Critical Synthesis of the Literature. *Latin American Perspectives* 1:4-29.

Chirot, Daniel. 1977. *Social Change in the Twentieth Century.* New York: Harcourt Brace Jovanovich.

Cochrane, Willard W. 1972. Agricultural Policy and National Development. In *Institution Building: A Model for Applied Social Change*, ed. D. Woods Thomas pp. 11-24. Cambridge, Mass.: Schenkman Publishing Co.

Consultative Group on International Agricultural Research. 1978. *Analysis by the TAC Review Team of Farming Systems Research at CIAT, IITA, ICRISAT and IRRI.* New York: CGIAR.

Cornet, Rene Jules. 1965. *Les Phares Verts.* Brussels: Editions L. Cuypers.

Crawford, J.G. 1977. The Future of the International System: A View from the Inside. In *Resource Allocation and Productivity in National and International Agricultural Research*, ed. Thomas M. Arndt, Dana Dalrymple, and Vernon Ruttan, pp. 590-98. Minneapolis: University of Minnesota Press.

Dahlberg, Kenneth A. 1978. An Evaluation of Research Strategies for Developing Appropriate Agricultural Systems and Technologies. Paper presented at meetings of the International Studies Association, Washington.

_____. 1979. *Beyond the Green Revolution.* New York: Plenum Press.

de Janvry, Alain. 1975. The Political Economy of Rural Development in Latin America. *American Journal of Agricultural Economics* 57:490-99.

Dillon, John L. 1976. The Economics of Systems Research. *Agricultural Systems* 1:5-22.

Dumont, Rene, and Nicholas Cohen. 1980. *The Growth of Hunger.* London: Marion Boyars Publishers.

Erisman, Alva Lewis. 1975. China: Agriculture in the 1970s. In *China: A Reassessment of the Economy,* pp. 324-29. Washington, D.C.: U.S. Congress Joint Economic Committee.

Esman, Milton J. 1972. The Elements of Institution Building. In *Institution Building and Development,* ed. Joseph W. Eaton, pp. 21-39. Beverly Hills, Calif.: Sage Publications.

Evenson, Robert E.; J.P. Houck; and V.W. Ruttan. 1970. Technical Change and Agricultural Trade: Three Examples: Sugarcane, Bananas and Rice. In *The Technology Factor in International Trade,* ed. R. Vernon, pp. 415-80. New York: National Bureau of Economic Research.

Evenson, Robert E., and Yoav Kislev. 1975. *Agricultural Research and Productivity.* New Haven, Conn.: Yale University Press.

Fishel, Walter, ed. 1971. *Resource Allocation in Agricultural Research.* Minneapolis: University of Minnesota Press.

Frankel, Francine. 1971. *India's Green Revolution: Economic Gains and Political Costs.* Princeton, N.J.: Princeton University Press.

Friedland, William H. 1974. Social Sleepwalkers. Davis: University of California Dept. of Behavioral Science, Monograph no. 13.

_____. 1979. Who Killed Rural Sociology? A Case Study in the Political Economy of Knowledge Production. Paper presented at the annual meetings of the American Sociological Association, Boston.

Friedmann, Harriet. 1978. World Market, State, and Family Farm: Social Bases of Household Production in the Era of Wage Labor. *Comparative Studies in Society and History* 20:545-86.

Gaitskell, A. 1959. *Gezira: A Story of Development in the Sudan.* London: Faber & Faber.

Gates, Paul W. 1960. *The Farmer's Age: Agriculture, 1815-1860.* New York: Harper & Row.

George, Susan. 1977. *How the Other Half Dies.* Montclair, N.J.: Allanheld, Osmun & Co.

Griffin, Keith. 1972. *The Green Revolution: An Economic Analysis.* Geneva: United Nations Research Institute for Social Development.

Hardin, Charles M. 1955. *Freedom in Agricultural Education.* Chicago: University of Chicago Press.

Harlan, Jack R. 1975. Our Vanishing Genetic Resources. In *Food: Politics, Economics, Nutrition, and Research,* ed. Philip H. Abelson, pp. 157-60. Washington, D.C.: American Association for the Advancement of Science.

Hatch, William Henry. 1886. Agricultural Experiment Stations. Report of the Committee on Agriculture, U.S. House of Representatives. Washington, D.C.: USGPO.

Hayami, Y. 1978. *Anatomy of a Peasant Economy.* Los Banos: International Rice Research Inst.

Heady, Earl O. 1971. Welfare Implications of Agricultural Research. In *Resource Allocation in Agricultural research,* ed. Walter Fishel, pp. 121-36. Minneapolis: University of Minnesota Press.

Hewitt de Alcantara, Cynthia. 1976. *Modernizing Mexican Agriculture.* Geneva: United Nations Research Institute for Social Development.

Institut International d'Agriculture. 1933. *Les Institutions d'experimentation agricole dans les pays temperes.* Rome: IIA.

_____. 1934a. *Les Institutions d'experimentation agricole dans les pays chauds.* Rome: IIA.

_____. 1934b. *Les Institutions de laiterie dans le Monde.* Rome: IIA.

International Association of Agricultural Librarians and Documentalists. 1966-70. *Quarterly Bulletin,* various issues.

Janzen, Daniel H. 1975. Tropical Agroecosystems. In *Food: Politics, Economics, Nutrition, and Research,* ed. Philip Abelson, pp. 103-10. Washington, D.C.: American Association for the Advancement of Science.

Kalitsi, E.A.K. 1970. The Organization of Resettlement. In *The Volta Resettlement Experience,* ed. R. Chambers, pp. 34-57. New York: Praeger Publishers.

Kirkendall, Richard S. 1966. *Social Scientists and Farm Politics in the Age of Roosevelt.* Columbia: University of Missouri Press.

Klein, Martin. A. 1979. Colonial Rule and Structural Change: The Case of Sine-Saloum. In *The Political Economy of Underdevelopment: Dependence in Senegal,* ed. Rita Cruise O'Brien, pp. 65-69. Beverly Hills, Calif.: Sage Publications.

Lowenthal, David. 1960. Population Contrasts in the Guianas. *Geographical Review* 50:41-58.

Masefield, G.B. 1972. *A History of the Colonial Agricultural Service.* London: Clarendon Press.

Moore, Ross E. 1943. Tingo Maria. *Agriculture in the Americas* (USDA) 3:107-8.

Moris, J.B. 1969. The Evaluation of Settlement Schemes Performance. In *Land Settlement and Rural Development in Eastern Africa,* ed. R. Apthorpe, pp. 79-102. Kampala: Transition Books.

Moseman, Albert H. 1970. *Building Agricultural Research Systems in the Developing Nations.* New York: Agricultural Development Council.

Palmer, Ingrid. 1972a. Food and the New Agricultural Technology. Geneva: UNRISD.

_____. 1972b. Science and Agricultural Production. Geneva: UNRISD.

Rogers, Everett, and Floyd Shoemaker. 1971. *Communication of Innovations.* New York: Free Press.

Rosenberg, Charles E. 1971. Science, Technology, and Economic Growth: The Case of the Agricultural Experiment Station Scientist, 1875-1914. *Agricultural History* 45:1-20.

Rostow, W.W. 1960. *The Stages of Economic Growth.* Cambridge: Cambridge University Press.

Russell, H.L. 1927 [1926]. Agricultural Education in the Orient and Australia. Proceedings of the 40th Annual Convention of the Association of Land-Grant Colleges. Washington, D.C., pp. 108-124.

Ryan, James G., and Hans P. Binswanger. 1979. Socioeconomic Constraints in the Semiarid Tropics and ICRISAT's Approach. Paper presented at International Symposium on Development and Transfer of Technology for Rainfed Agriculture and the SAT Farmer, ICRISAT, August.

Schultz, Theodore W. 1964. *Transforming Traditional Agriculture.* New Haven, Conn.: Yale University Press.

Schumacher, E.F. 1973. *Small is Beautiful.* New York: Harper & Row.

Seidel, Robert N. 1975. The Burden of Derivative Modernization. Paper presented at Rocky Mountain Council for Latin American Studies, Tempe, Arizona.

Sigurdson, Jon. 1977. *Rural Industrialization in China.* Cambridge, Mass.: Harvard University Press.

Sorrenson, M.P.K. 1968. *The Origins of European Settlement in Kenya.* Nairobi: Oxford University Press.

South Australia. 1887. Report of the Select Committee of the Legislative Council and the House of Assembly on Vegetable Production. Adelaide.

Spitz, P. 1975. Notes sur l'histoire des transferts de techniques dans le domaine de la production. Paris: Paper presented at a seminar on Science, Technology, and Development, OECD.

Sprague, G.F. 1975. Agriculture in China. In *Food: Politics, Economics, Nutrition, and Research,* ed. Philip Abelson, pp. 57-63. Washington, D.C.: American Association for the Advancement of Science.

True, A.C., and D.J. Crosby. 1904. Agricultural Experiment Stations in Foreign Countries. Washington: Office of Experiment Stations Bulletin 112, USDA, revised edition.

_____ . 1904. Needs of the Stations. U.S. Department of Agriculture, Office of Experiment Stations. Annual Report 1903:27-34.

United Nations. 1976. *Yearbook of International Trade Statistics, 1975.* New York: United Nations.

United States Department of Agriculture. 1903. Annual Report of the Office of Experiment Stations for the year ended June 30, 1902. Washington, D.C.: USGPO.

_____ . 1931. Workers in Subjects Pertaining to Agriculture in State Agricultural Colleges and Experiment Stations. Misc. Publ. 100. Washington, D.C.: USGPO.

Uphoff, Norman T.; John M. Cohen; and Arthur A. Goldsmith. 1979. Feasibility and Application of Rural Development Participation: A State-of-the Art Paper. Ithaca: Cornell University Rural Development Committee, Monograph no. 3.

Vallianatos, E.G. 1976. *Fear in the Countryside.* Cambridge, Mass.: Ballinger Publishing Company.

Waddimba, J. 1979. Some Participative Aspects of Programmes to Involve the Poor in Development. Geneva: United Nations Research institute for Social Development.

Wade, Nicholas. 1974. Green Revolution (I): A Just Technology, Often Unjust in Use. *Science* 186:1093-96.

Wallerstein, Immanuel. 1972. Three paths of National Development in 16th Century Europe. *Studies in Comparative International Development* 7:95-101.

_____ . 1974. *The Modern World System.* New York: Academic Press.

Wellhausen, E.J. 1975. Problems and Progress in the Acceleration of Maize Production in the Tropics. In *Agricultural Initiative in the Third World,* ed. the Agribusiness Council, pp. 53-65. Lexington, Mass.: D.C. Heath & Co.

7

Development in Sub-Saharan Africa and the Decline of Folk Knowledge in Agriculture

Thabo Fako

The decline of folk knowledge in sub-Saharan Africa is largely a result of contact with European settlers and colonizers. African contact with outsiders, however, is not a new phenomenon. In fact, as early as A.D. 710, some African states south of the Sahara maintained economic contact with Arabs. In 1415 the Portuguese captured Ceuta, a city in Morocco. In 1488 Bartholomew Diaz was blown south of the tip of Africa by a storm; he was followed in 1497 by Vasco da Gama.

As early as 1472, the French had already shown a special interest in the area of the Gambia and Senegal rivers. Through William Hawkins of Plymouth, the English in 1530 made their first contact with West Africa in the area now known as Sierra Leone. The Dutch West India Company, founded in 1621, soon exercised a virtual monopoly over the West African coast trade. In 1652 the Dutch established a colony in South Africa (Cape of Good Hope). The settlers, who were largely peasants, within a few years produced food and wine in sufficient quantity to become a valuable asset to the company.

Contact between European and African cultures has had far-reaching consequences that have interested scholars for a long time. By 1955 Fallers and others had observed that African studies were the home par excellence of structural sociological and social anthropological analysis, a tradition founded by Emile Durkheim, elaborated by Radcliffe-Brown, and more recently applied to empirical research by Fortes and Evans-Pritchard (Brady and Isaac, 1975:149). Much of this literature, however, lacks a critical edge. First, a large portion of it is administrative literature that looks at Africans as they ought to be, the way in which power among the natives ought to flow, and how native administration could be improved to enhance colonial interests

(Fallers, 1955). Second, the anthropological literature is preoccupied with how African societies *used* to be—i.e., how the traditional structures maintained their institutions in a harmonious and integrated fashion. Furthermore, both types of literature, although they appear to be "objective" interpretations, are conditioned by Western culture. Thus, in the final analysis, they are biased in favor of the colonial position.

These criticisms will not be elaborated upon here. The significance of these criticisms is that they permit an understanding of why folk knowledge declined, and why the corresponding bias toward Western science and technology in agriculture and the resulting social and economic problems in sub-Saharan Africa developed.

AFRICAN AGRICULTURE

Before devoting full attention to the social and economic problems of Africa, it is important to note the physical and geographic contours. Indeed, the social and the economic problems of Africa are difficult to separate from its physical, geographic, and historical problems. The continent is a large plateau with an elevation of about 1,000 feet above sea level. The plateau is higher in the South, where the elevation is over 3,000 feet. There are a few mountains, such as the volcanic Kilimanjaro in East Africa (19,321 feet), the Cameroon Mountains in West Africa (13,300 feet), the Atlas Mountains in Morocco, and the Drankesberg Mountains in the Southeast. There are no great gulfs or seas that extend deeply into the interior. While there are a few large rivers, they are generally not navigable year-round. In general the land is difficult to penetrate. (For a detailed discussion, see Cook, 1965; Hanse, 1964; and Sillery, 1961.)

Farmers are confronted by the problem of soil. Most of the continent lies within the tropics, and for several reasons good soil is a problem in this region. First, the high temperature of the soil speeds up the chemical processes that tend to destroy organic materials that enrich the soil. Second, warm water can hold more of a given compound than cold water. This, in turn, means that the tropical rainfall leaches the soil by carrying away the chemical elements necessary to plant growth. Third, water with a high mineral content is drawn to the surface during dry weather, and in time the soil forms a hard crust that contains little plant food (Cook, 1965:70). In addition to these problems soil erosion is especially active because much of the continent is a high plateau. With the recent introduction of the plow, much topsoil may be lost. This is especially true if plowing does not follow the contours of the land, or if fields are not separated by strips of uncleared land.

The problems of farming are further complicated by the fact that about a third of Africa has a tropical savanna type of climate that extends from the tropical rain forest to the margin of the deserts. The rainfall in this area is usually erratic from year to year, and the variations from season to season are extreme. Thus, while in a good year there may be sufficient rainfall to ensure an abundant harvest, the following year may produce famine conditions. Furthermore, the fact that much of the rainfall occurs in heavy showers makes farming difficult. Very little moisture penetrates the soil, and even this often does not last long enough to benefit the farmer.

The geographic and climatic conditions throughout much of Africa remain a problem not only to the indigenous farmer but to the modern agricultural scientist. The African farmer finds it difficult to protect himself against the hazardous conditions partly because he has relatively little indigenous knowledge regarding the construction of wells, irrigation, and the possibility of storing water for future use. The relatively sparse population discourages large-scale irrigation or terracing. As a result, the African farmer relies largely on the unpredictably patterned rainfall. In fact, rain is such a crucial factor in Africa that many customs and ceremonies practiced by Africans originated because of the need for rain. Among the Batswana peoples, for example, rain is so important that traditional doctors made elaborate preparations for it (Schapera, 1971). Compounds of many different ingredients were burned together in a potshed, crushed to powder, mixed with fat, and kept in a horn. This compound (*tshitlo*) was made from the roots of several plants, the body of a frog (*segogwane*) or a small rain frog (*senanatswii*), portions of a lightning bird, and, whenever available, skin from the chest and belly of a crocodile (*kwena*). According to Schapera, the frog was chosen because it is an animal associated with water, the lightning bird because it "brings the rainbearing clouds from the sea," and the crocodile because it was very good medicine for attracting rain (Schapera, 1971:49).

Because of the many problems that appear to oppose the agricultural development and health of the African peoples, Africa has had a reputation as the "dark" continent. Indeed, Africa is more heavily parasitized than any other continent (Thomas, 1965). The prevalence of disease carried by mosquitoes, tsetse flies, and locusts, the absence of an adequate transportation system, the problems of soil and water, and climatic problems make it difficult at best for the indigenous or the Western agriculturist to practice successful agriculture in Africa.

The difficulty of life in Africa was even more evident to early European settlers. The high death rate among Europeans who ventured into Africa in the early years, especially before the development of quinine, appears to support this notion.

Despite the above mentioned problems, African development is synonymous with African agricultural development (Brietze, 1976). The fact is that most African countries are rural. In Ethiopia, for example, 92 percent of the people generate 60-65 percent of the gross domestic product and 90 percent of the exports through agriculture, the only significant economic activity pursued in rural areas (Brietze, 1976:637). This is also true in Tanzania, where 95 percent of the people live in rural areas. The industrial sector in Tanzania contributes only 10 percent to its total monetary income (Desai, 1976).

Significant socio-economic changes in African colonies were not evident until the end of the nineteenth century, when a new kind of subsistence culture was practiced; and maize, millet, and sorghum became the main crops cultivated (Dixon-Fyle, 1977:579). Since the beginning of the 20th century, however, subsistence farming and the network of relationships surrounding it have been rapidly losing ground. In Rhodesia the early European settlers (who were primarily maize farmers) introduced the plow in the 1920s (Dixon-Fyle, 1977:596). Hoe farming had allowed for shifting cultivation but had not developed incentives for continuous cultivation of one piece of land. In shifting cultivation, once a piece of land begins to lose its fertility it is left in search of more fertile land. Plowing, on the other hand, meant a more permanent occupation of land with the corresponding ownership rights.

As the plow and other farm implements were brought into use on African soil, the institutions of a capitalist economy were created and imposed upon the traditional subsistence economies (Fallers, 1955). Such changes were experienced by Uganda in the decade from 1900 to 1910. Similar changes occured in Kenya during the 1950s and early 1960s (Benard, 1972). These rapid changes not only resulted in a market economy in which cash crops such as coffee, tea, tobacco, and cotton became increasingly important, but also reshaped the way of life of the people undergoing these changes.

The colonial administration brought with it not only changes in agriculture, but changes in the law of the land as well. Since customary law appeared unconducive to economic development, a new system of law that would promote economic development was introduced. Land owned by a group in the customary sense was not accepted by banks and other moneylenders as security for loans. When the freehold system was introduced in English colonies, the customary land-tenure system, however, was not abolished. Thus, the African could occupy land under two systems of law. But, as Kiapi (1975) argues, a system of law that places one leg of a nation in the past and one in the modern age is not conducive to development.

Since the 1970s, emphasis on economic development by African governments has meant seeking every way to improve existing technologies in order to increase productivity. This does not necessarily require Western expertise. On the other hand, there is no need for Africans to rediscover fundamental scientific laws or to reinvent the techniques of applying them. Thus, it has been argued that the technological knowledge of the Western world can quickly be transferred to new countries (Foster, 1962). The principle behind this notion is that the administrative, agricultural, health, and educational skills needed to satisfy the new needs of African nations can easily be taught. With help from wealthier nations, African countries should be able eventually to develop their own resources to a degree adequate for their needs. In the light of the above, the spread of Western education has caught on like wildfire as each African state develops its own manpower (Ofuatey, 1976:229).

The commitment of the African leaders to educational expansion is not only a response to popular demand but also a reflection of an ideology of social change that posits that (Western-style) education is the preeminent instrument for promoting desirable social and economic changes (Foster, 1962: 183). The immediate result of this emphasis on Western-style education is that folk knowledge is neglected. The long-term result is that folk knowledge and traditional methods are being discarded.

It is not surprising that African nations promote Western-style education. African societies have had centuries of contact with Westerners who established schools that provided clerks for the civil service, catechists and teachers for the missions (Lloyd, 1968), and manpower for many other colonial jobs. At independence, power passed to a Western-educated elite that continued the use of English or French as the official language. Thus, since the innovating group that is expected to lead in the development of new values (Court, 1976) has been socialized to accept Western technology and to a considerable extent its value system, a Western bias toward development—what Lipton (1977) has called "urban bias"—appears inevitable.

The group that is expected to lead in reflecting the values of its society and, if necessary, in rejecting Western influences is the very group that is effecting the decline of its own traditions. Thus, paradoxically, much for which the colonial governments were blamed continues under new African governments.

The decline of folk knowledge and the corresponding and overwhelming Westernization seems to be independent of leadership types. Indeed, leadership types may affect the rate and extent of processes; but the decline of folk knowledge itself must be sought elsewhere than in the conscious intentionality of those who are ruling. Agricultural problems, like other social problems, originate in a redistribution of population accompanying a rapid urbanization and a reorientation of production toward a market economy (Green and Fair, 1965). In all cases, the social and economic structures are becoming more complex, and the functions performed more numerous as the cultural and material techniques of Western civilization are absorbed and adapted by African nations. Therefore, it can be said that the absorption and adaptation of techniques, while reflecting ideological allegiances, does not prove positive or negative intent; nor does the intent in itself reduce or create problems as such. For example, it has been argued that, as early as the late 1940s, the colonial governments of sub-Saharan African nations began to show more positive interest in the agricultural development of those nations (Schatz, 1972:128). However, it is clear that their "positive" intent has not reduced agricultural problems in those nations.

By encouraging Western-style technological advancement for their nations, African governments have nourished conditions that favor the decline of folk knowledge in agriculture. Like their Western advisers, they recognize that the facts of food and population in the world today and the evident poverty of most of their people suggest that agricultural science and technology, although they have much to offer, are not being sufficiently applied in practice by the vast majority of farmers (Leagans and Loomis, 1971:439). Because of this, they promote agricultural research that aims to discover how the land may be made to yield more of the commodities required for the life of the people and for trade (Bohnet and Reichelt, 1972:44). As this is done, methods such as shifting cultivation and hoe farming are being replaced by new technologies. The aim of agriculture has become the development of techniques that yield more food on less land.

MODERNIZATION AND AGRICULTURAL DEVELOPMENT

The agricultural extension worker in Africa, as elsewhere, attempts to alter traditional ways by demonstrating the advantages of the new (Spicer, 1952). In most cases the decision that the present system in a locality is "wrong" has already been made. The "traditional" is already defined as inferior, and research work is done to demonstrate this view. By definition, what the extension worker or change agent has to offer must be better than what already exists.

Farming problems are usually analyzed not in terms of local beliefs and mythology but as a continuous interaction of soils, climate, microorganisms, crops, pastures, natural plant and animal communities, livestock, markets, costs, and man and his environment (Nsubuga, 1976:1). The emphasis, as Nsubuga argues, is on the fundamental biology and economics of this complex. Correct understanding of principles involved in growing the plants is also heavily emphasized (Opare, 1977:80-81).

Agricultural problems are seen by change agents in terms of the fact that in many regions agricultural science and technology are not appropriately developed. Furthermore, it appears that agricultural scientists and technologists do not sufficiently understand their role in the process of change (Leagans and Loomis, 1971).

Development or modernization is seen as the process of change toward those types of social, economic, and political systems that have developed in Western Europe and North America from the seventeenth century to the nineteenth and have then spread to other countries (Brett, 1973:3). Agricultural development or modernization is evaluated by its proximity to institutions and values of Western and particularly Anglo-American societies (Tipps, 1973). By deriving the attributes of "modernity" from a generalized image of Western society and then positing the acquisition of these attributes as the criterion of modernization, agricultural researchers have attempted to duplicate the West in Africa. However,

> Western models can only provide a point of reference, they cannot be transplanted unmodified from one society to another. Therefore, societies wishing to make use of Western experience and technology will only be able to do so if they themselves have an adequate theoretical grasp of their own situation—of their needs and resources, and the possibilities for the mobilization of their populations. . . . The new technology will make a positive contribution, only where it is relevant to local needs

and can be adapted to suit local conditions. . . . This means that the understanding of Western modes must be backed by an equally profound understanding of the nature of developing societies—unless this occurs the result will be the creation of expensive but useless monuments to Western technology. . . . Carried out in this way the adoption of Western models is not an imitative and dependent activity; it is a creative process which must ultimately result in the evolution of new social forms and autonomous models of thought (Brett, 1973:16).

This is correct as far as it goes, but a thorough understanding of the cultural forms and values of the people of another nation is not in itself sufficient to eliminate human problems (Foster, 1962:6).

Development is a human activity and not an exclusively scientific one. By leaving out value judgments, technology can tell us absolutely nothing more than that "for the attainment of a given technical end X, Y is an appropriate means." The "real world" is a world in which objective as well as subjective criteria, and not either alone, are employed (Weber, 1949). Purely "objective" statements are as dangerous as purely "subjective" ones. In this light the appreciation of agricultural systems *and* traditional folk belief systems cannot be overemphasized. Appreciation of traditional agricultural systems is, according to Saint and Coward (1977:735), a necessary first step before any planned intervention is attempted.

Tradition is a source of truth. It contains only that knowledge which has been verified by a (fallible) community (Busch, 1978). According to Gusfield (1967), the new and the old can be and frequently are mutually reinforcing, rather than opposing. Very often, the desire to be modern and the desire to preserve tradition operate simultaneously. Nationalism, for instance, is deeply committed to both horns of the dilemma of tradition/modernity (Gusfield, 1967:359). Furthermore, it is important to note that we cannot easily separate the terms "modernity" and "tradition" from some *specific* tradition and some *specific* modernity. The modern comes to the traditional society as a particular culture with its own traditions (Gusfield, 1967:361). The Western tradition comes to African nations and replaces their local traditions. Thus modernization and Westernization in this respect may be used synonymously.

Insofar as the theme of agricultural development or modernization is that of seeking to alter a people's way of life and of "improving the well-being and happiness of generations of men and women" (Spicer, 1952), the administrators of a program of agricultural change carry a heavy responsibility. Their efforts to increase productivity will depend considerably on the success with which agrotechnologies and appropriate organizational and institutional arrangements can be incorporated into systems of traditional agriculture, especially those in which small farmers predominate (Saint and Coward, 1977:733). The capacity to develop technology consistent with physical and cultural endowments is the single most important variable accounting for differences in agricultural productivity among nations (Arndt, Dalrymple, and Ruttan, 1977:3).

Despite the "highest priority" assigned to rural development in the policy documents of many African governments, neither the allocation of public funds nor the implementation of development strategies has been energetically directed toward improving the living standards of the rural masses (Ollawa, 1977:401). Equally, the recognition of the need for a new development model by an increasing number of scholars has yet to produce adequate guidance for those African regimes which are seeking to transform the rural areas. The result of this is that there continues to be a great disparity between the promulgated goals of development and their actual results. Frequently, the strategies adopted do not relate to the local political economy or to the structural capability and social orientation of the environment (Ollawa, 1977:402).

The ultimate result of agricultural development and modernization to this point has been to encourage the promotion of policies that bias resource allocation and income distribution in favor of foreign investors and the affluent elite, including the political and administrative ruling class, and favor the urban centers at the expense of the rural areas (Ollawa, 1977). This condition is indeed regrettable. It must be pointed out, however, that even among critics there is a general agreement that whatever else it may be, modernization is a type of social change that is both transformational in its impact and progressive in its effects (Tipps, 1973:202). Furthermore, it is a result of the expansion of man's rational control over his physical, social, and psychological environment (Hall, 1965:21). In addition, becoming modern means being "progressive," and tradition acts as a barrier to "progress."

> Modernity assumes that local ties and parochial perspectives give way to universal commitments and cosmopolitan attitudes; that the truths of utility, calculation, and science take precedence over those of emotions, the sacred, and the non-rational: ... that identity be chosen and achieved, not ascribed and affirmed; that work be separated from family, residence and community in bureaucratic organizations; that manhood be delayed while youth prepares for its tasks and responsibilities (Tipps, 1973:20).

To the extent that the major direction of social change today is from the "parochial" to the "cosmopolitan" forms of social organization and from popular to scientific subsystem value orientations (Suchman, 1965:16), folk knowledge in agriculture, as in any other field, is bound to decline in use. If this judgment is correct, then, a major problem of modern times will be to reduce the gap between a rapidly advancing scientific technology and a hesitant status quo. This can probably be done more successfully by seeking new organizational forms and modes of operation that, while taking advantage of scientific advances, still are translatable into terms acceptable to the "parochial" segments of the public. It makes no sense to talk of "calling a halt" to scientific progress, but much can be done to frame such progress so that it is less foreign to traditional forms and methods. Similarly, as Suchman

(1965:16) suggests, while insofar as possible people need to be educated to this new scientific approach, they cannot be threatened or exhorted to change. The major share of responsibility must remain with the changing social institutions to seek ways of reaching people with new ideas and techniques fitted as closely as possible into old and traditional clothing.

In conclusion, we have suggested that the bias against folk knowledge is not a simple fact or the result of a conspiracy. Rather, it is the result of efforts by African governments to combat agricultural and health problems associated with climate, soil, and natural and human resources. Furthermore, we have argued that agricultural scientists often do not take folk knowledge seriously. As a result, they make unnecessary, and occasionally tragic, mistakes. Moreover, as folk knowledge is replaced by "modern" methods, a great deal is lost. Folk practices are based upon generations of experience; even when they cannot be "explained" by the practitioners of science, they may nevertheless work.

REFERENCES

Arndt, Thomas M.; Dana Dalrymple; and Vernon Ruttan, eds. 1977. *Resource Allocation and Productivity in National and International Agricultural Research.* Minneapolis: University of Minnesota Press.

Benard, Edward F. 1972. *East of Mount Kenya: Meru Agriculture in Transition.* New York: Humanities Press.

Bohnet, Michael, and Hans Reichelt. 1972. *Applied Research and Its Impact on Economic Development: The East African Case.* Munich: Weltform Verlag.

Brady, Ivan A., and Barry L. Isaac. 1975. *A Reader in Culture Change: Volume II Case Studies.* New York: John Wiley & Sons.

Brett, E.A. 1973. *Colonialism and Underdevelopment in East Africa: The Politics of Economic Change 1919-1939.* New York: N.O.K. Publishers.

Brietze, Paul. 1976. Land Reform in Revolutionary Ethiopia. *Journal of Modern African Studies* 14:637-60.

Busch, Lawrence. 1978. Understanding Understanding: Two Views of Communication. *Rural Sociology* 43:450-74.

Cook, Arthur N. 1965. *Africa: Past and Present.* Totowa, N.J.: Littlefield, Adams & Company.

Court, David. 1976. The Educational System as a Response to Inequality in Tanzania and Kenya. *Journal of Modern African Studies* 14:661-90.

Desai, Priya V. Mutalik. 1976. Ujamma Villages: A Tanzanian Experiment in Rural Development. *African Quarterly* 16:36-55.

Dixon-Fyle, Mac. 1977. Agricultural Improvement and Political Protest on the Tonga Plateau, Northern Rhodesia. *Journal of African History* 18:579-96.

Fallers, Lloyd A. 1955. The Predicament of the Modern African Chief: An Instance from Uganda. *American Anthropologist* 57:290-305.

Foster, George M. 1962. *Traditional Cultures and the Impact of Technological Change.* New York: Harper & Row.

Gusfield, Joseph R. 1967. Tradition and Modernity: Misplaced Polarities in the Study of Social Change. *American Journal of Sociology* 72:351-62.

Green, L.P., and T.J.D. Fair. 1965. *Development in Africa: A Study in Regional Analysis with Special Reference to South Africa.* Johannesburg: Witwatersrand University Press.

Hall, John W. 1965. Changing Conceptions of Modernization of Japan. In *Changing Japanese Attitudes toward Modernization,* ed. Marius B. Jansen, pp. 7-42. Princeton, N.J.: Princeton University Press.

Hanse, William A. 1964. *The Geography of Modern Africa.* New York: Columbia University Press.

Kiapi, Abraham. 1975. Legal Obstacles to Rural Development in Colonial Uganda. *Mawazo* 4:101-11.

Leagans, Paul J., and Charles Loomis. 1971. *Behavioral Change in Agriculture: Concepts and Strategies for Influencing Transition.* Ithaca, N.Y.: Cornell University Press.

Lipton, Michael. 1977. *Why Poor People Stay Poor: Urban Bias in World Development.* Cambridge, Mass.: Harvard University Press.

Lloyd, P.C. 1968. *Africa in Social Change: Changing Traditional Societies in the Modern World.* New York: Penguin Books.

Nsubuga, H.K. 1976. Professional Needs in Agriculture in Uganda. *Uganda Journal* 38:1-10.

Ofuatey, Kodjoe W. 1976. Education and Social Change in Africa: Some Proposals. *Journal of African Studies* 3:229-46.

Ollawa, Patrik E. 1977. On a Dynamic Model for Rural Development in Africa. *Journal of Modern African Studies* 15:401-23.

Opare, Dua K. 1977. The Role of Agricultural Extension in the Adoption of Innovations by Cocoa Growers in Ghana. *Rural Sociology* 42:72-81.

Saint, William S., and Walter E. Coward, Jr. 1977. Agriculture and Behavioral Science. Emerging Orientations. *Science* 197:733-37.

Schapera, Issac. 1971. *Rainmaking rites of Tswana tribes.* Cambridge: African Studies Center.

Schatz, Sayre P. 1972. *South of the Sahara: Development in African Economies.* Philadelphia: Temple University Press.

Sillery, Anthony. 1961. *Africa: A Social Geography.* London: Duckwork.

Spicer, Edward H. 1952. *Human Problems in Technological Change: A Casebook.* New York: John Wiley & Sons.

Suchman, Edward A. 1965. Social Patterns of Illness and Medical Care. *Journal of Health and Human Behavior* 6:2-16.

Thomas, John D. 1965. Some Preliminary Observations on the Ecology of a Small Manmade Lake in Tropical Africa. In *Ecology and Economic Development in Tropical Africa,* ed. David Brokensha, pp. 113-46. Berkeley: University of California Press.

Tipps, Dean C. 1973. Modernization Theory and the Comparative Study of Societies: A Critical Perspective. *Comparative Studies in Society and History* 18:199-225.

Weber, Max. 1949. *The Methodology of the Social Sciences.* Translated and edited by Edward Shils and Henry A. Finch. New York: Free Press.

8

Four Decisions
Facing Latin American Extension

Donald K. Kazee

A survey of literature on agricultural extension in Latin America reveals that there is considerable debate as to the nature of the mission of the *extensionista* and how he is to accomplish that mission. These differences in opinion in the Latin American extension community are no mere academic exercises; they are fundamental choices that will determine whether Latin America will be able to cope with its burgeoning population, or whether its best efforts will be swept away by a tide of human misery.

The controversies revolving around extension in Latin America center on three basic questions: What is the goal of extension? How is extension to be achieved? To whom are extension efforts directed? And because these questions are reflective of disputes within the North American agricultural community from which many Latin American extension services sprang, one might ask: To what extent has American influence been a beneficial force in Latin American extension?

EXTENSION FOR WHAT?

One can scarcely begin to discuss the nature of extension unless one almost casually commits himself to one side or the other of these basic divisions, if only by the language he chooses. However, I will endeavor to formulate a fairly neutral definition to use as a point of departure. The word "extension" implies a transfer or an expansion. There is a debate as to exactly what is to be extended, but for now it is sufficient to refer to the extension of agricultural innovations. If, then, we are extending agricultural innovations, to what end

My appreciation to Dr. Louis Quiros Varela for his assistance in preparing this chapter.

does this extension take place? There are two opposing viewpoints on the answer to this question. One avers that the purpose of agricultural extension is to improve the quality of life of those who make their living from the land— i.e., the rural population. The other viewpoint holds to a stricter definition of agriculture and maintains that increased agricultural production ought to be the goal of the spread of new ideas.

The "people-oriented" (rather than "production-oriented") view was put forth in Chiclayo, Peru, in 1970 in the recommendations of the Technical Conference on Agricultural Extension and Rural Youth. The Conference recommended that "the basic objective of rural extension service be established as the development of the human being, taking as the center of action the rural family in its integral form, utilizing a familial atmosphere as a medium of education in accordance with the interests and needs of these groups" (FAO, 1971:24). This broad view of extension is echoed by Juan Galecio Gomez of the FAO mission in Ecuador: "Agricultural extension is a program of education with the aim of aiding the population in the search for solutions to its economic, social, and cultural problems, so that they may live a better life by means of the rational use of their human, physical, and economic resources" (Lopez Cordovez and Galecio Gomez, 1961:7).

Obviously, this is quite an awesome task to put into the hands of the lone *extensionista* laboring, for example, in the Andean highlands. However, as long as the extension services keep this broad, long-range goal in mind, they will not, according to the holders of this "humanist" viewpoint, be led down the primrose path of production promoting. Implicit in this view of agricultural extension are a whole gamut of services that are not strictly agricultural in the "fertilizer, maize, and tractor" sense of the word.

Strictly agricultural services must be augmented by other services which will aid in the amelioration of the life of the *agricultor*. What is to be desired is not merely a more efficient producer, or a *campesino* who uses an innovation for a while without an understanding of why it will improve his lot, but an individual who has a new frame of mind regarding innovation and his role in society. The extension service should foster a state of mind in the *campesino* in which he is receptive to innovations generally and is cognizant of the benefits that new means of production can bring his family. The *campesino* who uses an innovation without understanding it will have little incentive to try anything else. However, a basic understanding of the principle of innovation and of the possible rewards can motivate the *campesino* not only to use more innovations himself but to provide an endorsement of the innovations for his neighbors. This state of mind and trust in the extension agent can be fostered through information on health, sanitation, maternity, nutrition, food storage where appropriate, the market and monetary system so far as it applies to him, home building, and literacy skills both in his own language and in Spanish or Portuguese. If any one area is to be emphasized over the other, it is literacy, which should take precedence because it is the key to more effective learning in other areas. Education, not only agricultural but basic, will provide a solid foundation for solid gains in the future. With education, it is hoped the farmer will perceive a somewhat brighter future in rural life and not be tempted to flee to the city, only to become a shantytown dweller.

This model of extension, although an optimistic one in theory, has several limitations, not the least of which is manpower. This model of extension requires not only expert agronomists but teachers, health-care personnel, carpenters, nutritionists, and the like. There is simply not enough personnel to provide an extension service such as this one to every village south of the Rio Grande. Even if one restricted the number of villages to be served, there is a trade-off between the number of people who can be served minimally and the degree to which a limited number can be served with a semblance of comprehensiveness.

A second limiting factor would be the cost involved. So extensive a program would certainly be restricted by the already overblown public expenditures in countries plagued by galloping inflation. The same trade-off as in personnel would have to be made regarding quantity versus quality.

A third limitation would be the political reality. The problem with increased awareness of agricultural innovation, increased literacy, and increased material expectations is that these types of awareness can scarcely be separated from political awareness. Farmers who can put agricultural innovations to work just might want enough land to make them work better. The peon who gives his landlord a major part of his crop for the use of precious little land just might decide that his crop could bring more in a monetized market. He also might decide that he could buy more and in more variety on the monetized market than in the cozy "company-store" arrangement he has with his landlord in which payment is made in kind. The *latifundista* does not know for sure but strongly suspects that a community of literate, socially aware *campesinos* with high expectations might try to pull off one of these tricks. If pressure became strong enough nationwide, they might even force through a program of agrarian reform, which "everyone" knows is communist-inspired and a threat to hemispheric security. If extension is perceived as too effective, *extensionistas* might have trouble gaining access to some of the *campesinos*.

The other view of extension in Latin America is production-oriented. Extension is not to benefit any group in particular but, by increasing production, to benefit the nation as a whole. AID Policy Coordinator Edward Rice, in his study of Andean extension, declares that "increased productivity and production of marketable goods ought to have been the principal objective [of extension], since remunerative employment seems to be the most powerful engine of a development process that, properly managed, offers to achieve all these [humanistic] objectives" (Rice, 1974:157). Production-oriented extension is also desirable in the macroeconomic sense, argues Roberto Alemann of the Union Bank of Switzerland in Argentina. That the family unit is favored . . . as a necessary form of agricultural enterprise . . . hark[s] back to the European Middle Ages, when possession of land signified freedom from the feudal lord. But the only important thing today is whether the enterprise is efficient, whether it is a family unit or not. . . . [To] refer disparagingly to the sizes of both the microfarm and the giant estate [is to] take no account of what really matters, that is, their efficiency regardless of size or organization. . . . What is desired (and this is the objective of the policy of promoting agriculture) is that the rise of agricultural production be considerably steeper than the growth of the population in Latin America during the next decade. There will then be sufficient production to improve both the caloric content and the composition of the diet, to improve the living standards of both the rural and the urban population, and with more exports of agricultural products, to earn foreign exchange to finance general economic development including agriculture (IDB, 1967:211-13).

If one makes certain assumptions, this strategy is surely one that makes sense as a role extension can play in long-range development. Even if certain social goals have to be sacrificed in the short run, extension with an emphasis on production will in the long run result not only in increased caloric consumption for the *campesino*, but also in capital for industrialization, using agricultural surpluses as export crops. This industrialization would absorb the rural workers who had previously fled the countryside for the cities.

However, the assumptions upon which this overall strategy for the role of extension is based are not realistic. First, mere emphasis on agricultural production does not mean an increase in caloric consumption. There is a limit to the amount of bananas one can consume, and coffee is not noted for high caloric value. An increase in the production of these and other crops grown mainly for export will not be a panacea for diet deficiencies. If agriculture is continued with current *latifundistas* controlling means of production, more profitable export crops will be grown instead of consumption crops.

Second, even if one granted that food crops would be produced, the assumption that these food crops could be exported is naive. The market for such crops is other underdeveloped countries who can ill afford to import quantities of food crops on a par with America's ability to import coffee. Hard currency needed to finance industrialization would not be forthcoming from nations with a need for food imports. Too, a Latin American grain industry would run into head-to-head competition with such agricultural powers as the United States, Canada, and Australia.

Third, the productionists quoted above assume a fully functioning monetized market system that would distribute remuneration to those who grow these greater amounts. Again, they have paid no attention to issues relating to current control of the means of production and do not anticipate the cut in income that landlords may demand of *campesinos*. Under current land control, it is the *latifundista* who will accrue remuneration, not the *campesino*. Productionists have assumed that, should the peasant receive remuneration, he will have access to some market other than the "company store" controlled by the *latifundista*. It is assumed that there are markets that will convey to the *campesino* his needs.

Fourth, it is assumed that governments will use export earnings to reinvest in agriculture. The current urban political power centers are not recognized (Lipton, 1977). Fifth, it is assumed that investment of export earnings in industrialization will be carried forth prudently and will not be used for showboating and grandstanding by the current junta. Sixth, it is assumed that there will be investment in labor-intensive means of production to absorb unemployment in the city and the countryside rather than the capital-intensive means of production currently being advertised by industrialized countries with technology to export.

Seventh, it is assumed by the productionists that their definition of "development" is adequate to point out solutions for the problems of underdevelopment. Social problems integral to the nature of underdevelopment are simply not defined within the scope of the solution. Problems which require solutions that are not measured in units other than bushels are ignored. Others, however, would expand this narrow definition of development. It is pointed out by Jose Marull, Director of Planning for the Inter-American Institute of Agricultural Science, that development (and therefore extension as a process of development) is to pursue

> the elevation of human dignity, by means of satisfying the basic needs of the individual . . . , [those being] purpose in life, a sense of belonging to a social group, access to creative opportunities, the enjoyment of goal fulfillment, to breathe freely, to share power and develop in an atmosphere of personal security manifested in sufficient food, clothing, housing, education, recreation and health, the protection of authority, and acceptable work (IICA, 1965:91).

Mere calorie counting falls short of fulfilling these human requirements. Production for the sake of production, in the judgment of this writer, is an invalid goal. If one assumes, as Lappe and Collins (1977) seek to demonstrate, that inadequacies of food are due not to deficiencies in overall supply but to the distribution of that supply via current economic systems, an increase in production would do nothing to solve the problems of inadequate living conditions.

On the other hand, production as a means of achieving more humanistic ends is an entirely acceptable goal. As Monsignor Luigi Ligutt, Observer of the Holy See at the Food and Agricultural Organization in Rome, perceives, "A farmer whose per hectare and per hour productivity is satisfactory or even superior is not merely enjoying a greater income, but, by the very process of higher productivity, his own total personality is improved, consciously or unconsciously, his abilities have brought him out and he has prepared himself for more effective future actions" (IDB, 1967:197). It must be emphasized here that it is productivity per man, not productivity per acre or per *latifundia,* that yields these benefits; and it is to increasing this type of productivity that extension must be directed. Deficiencies in trained personnel and money to finance extension services do not detract from the appropriateness of the human-centered approach to extension. They only extend the time needed to achieve the goals of extension.

HOW IS EXTENSION TO BE ACHIEVED?

If one accepts that extension is to help the rural man to "breathe freely" and share in the power to make decisions in his own life, one is brought to the second fundamental question which divides Latin American agriculture: How is extension to be achieved? This question is concerned with the ambiguity previously referred to in the term "agricultural innovation." The form in which "agricultural innovations" are to be delivered determines what method of extension is to be used. Is agricultural innovation to be presented merely as information? The extension agent makes known to the farmer that a certain innovation exists, that it has worked in the past, that there is a reason it works, and that it will work for him. The agent asks if the farmer would like to try it. Utilizing his free will and power over his own life, the *campesino* may say yes or no. If the *campesino* agrees to try, there is no problem. If he refuses, one must ask: "Has extension been achieved?" As in the case of the *campesino,* the answer may be yes or no. It may also encompass both. Yes, the *campesino* has exercised his own free will over his means of production. Information has been presented and the farmer has used his best professional judgment.

But also, no. True, he has used his will over his means of production; but what he has not utilized is important. He has not utilized the agent's perspective of the state of the practice of agriculture. He has not utilized a time reference frame much larger than the immediate future. He is not thinking of the possibilities for this and corollary innovations five or ten years from now. He has not utilized a geographic reference frame any larger than his village. He does not know that with the innovation he can invest less and get more to supply a demand perhaps hundreds of miles away. He has not utilized a possibilities reference frame any broader than the immediate agricultural consequences. He does not think of building a better home with the money to be made from this innovation. Assuming the extension agent has something that merits some consideration (and this cannot be assumed in every case), it must be concluded that the *campesino* has in fact limited his future freedom of will. Extension has not been achieved. The *campesino* lacks the criteria by which a proper evaluation of the agent's offering could be made.

It is the "si" response that denotes the *educativa* school of thought in Latin American extension. In the tradition of being servants of the people, the *educadores* make every effort to let those they serve determine the direction of the extension effort. *Campesinos* are surveyed to establish priorities, and the extension agent seeks to show them how to fulfill their most pressing needs. Information is presented in an unbiased, nonpromotional way. This method works very well for Iowa farmers who love government when government has what they want when they want it, but who loathe to have Washington push anything down their throats. Indeed, the *educativa* school in Latin America began when Americans began to help found the extension services in the republics. These ideas of the farmer's self-direction were based on the backgrounds of former Federal Extension Service agents working in Latin America who believed that no purpose was served in doing the farmer's job for him (Rice, 1974:62). If the innovation was to be adopted with lasting effect, the farmer had to decide for himself that it ought to be done.

However, one first has to know that there is a choice before the decision can be made. One has to know what questions to ask. Edward Rice describes the problems of farmer-directed extension, or Program Building, in Ecuador:

> The difficulty of Program Building is that subsistence farmers can neither identify nor solve their real problems. . . . It is inconceivable that village elders in Asia or Latin America would originate a request for short, stiff-stalked, high-response wheat. What they are more likely to ask for . . . are more park benches and shoes. These are the sort of problems which they can hope to solve. The other solutions are not conceivable. The farmers and their leaders are no more likely to ask for a dwarf wheat, as a solution to their problem with lodging, as for the wind to stop blowing. There is no denying that the small farmers of Latin America are rational, wise people . . . but they don't know which constraints can be relaxed by purposeful human intervention and which cannot (Rice, 1974:127).

Those who recognize this difficulty and urge stronger intervention would answer no—that extension in the previous example had not been achieved. They form the *fomento* school of extension. *Fomentar,* logically enough, means to foment, to promote. It is a shift away from making innovations available to making innovations attractive. *Fomento* may involve propaganda, large demonstration farms, glamorization of the innovation, extension of credit to finance the innovation, free distribution of equipment, or anything else to entice the farmer to use the innovation.

Fomento has an advantage over the *educativa* method in that it provides a motivation for the *campesino* that strict education does not. W.F. Whyte, in studying an extension program in Peru, notes that "as individuals pursue a course of action and find themselves rewarded (psychologically reinforced) by the consequences of their actions, they become more strongly committed to that course of action" (1977:22). Rice elaborates on this concept when he says that *campesinos* "have to get involved in the new program, or observe a neighbor who is involved" (1974:127). The idea is that once a peasant makes a trial commitment to a new idea, he is more likely to make a larger commitment.

Fomento had its beginnings in the early days of the Department of Agriculture in the United States, and before World War II in Latin America, when in both instances seed and fertilizer were distributed to the farmer or were offered at reduced prices. The problem was that in Latin America the farmer might or might not know what to do with it (IICA, 1965:32). With the flood of U.S. influence in the 1940s, the *educativa* school became *de modo*. However, since the mid-sixties, *fomento* has made a comeback (Rice, 1974:124). American influence on the controversy was not entirely *educativa*. Latin American agriculturalists were trained by both Federal Extension Service personnel and pro-*fomento* Soil Conservation Service personnel, and Latin America inherited a feud that had been brewing in the United States since the 1930s (Rice, 1974:62).

What are the criticisms of *fomento*? First, the old Federal Extension Service argued that it was to no avail to try to bulldoze farmers into the twentieth century. Second, it might be possible for someone with an axe to grind to use the machinery of *fomento* to push products or practices in which he had a vested interest. This would not only be unfair to the *campesino* but would erode his trust in other valid campaigns brought forth by the government. Third, the effects of a short-term campaign, no matter how intense, are not likely to produce any permanent change in the *campesino*'s way of doing things without a program of basic education.

To this observer, the *"educativa-fomento"* dispute is a false dichotomy. *Fomento* and basic information extension are mutually supportive activities. Simply knowing about crop rotation does no good unless one has more than one kind of seed to plant, and it does little good to plant different types of seed at random if one does not know how or why or whether to plant the different types. The major disadvantage of both schools of thought is the time wasted in argument over the correct approach. This view is affirmed by Fernando del Rio of the University of Puerto Rico:

> *Fomento* has positive elements which can be utilized advantageously by the extension agent. Among these are the concentration of efforts in determined areas, technical assessment in large or small degree of new or improved practices, and the facilitation of means by which activities can be realized. Extension [*educativa*], on the other hand, can complement the work of the *fomentista,* reinforcing and consolidating the educative aspects (IICA, 1965:33).

It is not a matter of pursuing one method at the expense of the other. There are many roads to development, and these two means of extension can complement each other so that in the end the *campesino* can truly exercise his free will.

EXTENSION FOR WHOM?

The third central question facing the Latin American extension community is "To whom should extension efforts be directed?" The issue here is one of comparing the costs of extension with the benefits. One opinion is that scarce extension personnel and money should be directed to those best able to use the extension service. This would include the current rural upper and upper middle classes who already are endowed with land, literacy, and a receptiveness to innovation. Direct return on extension investment would theoretically be immediately forthcoming in terms of increased production.

The other school of thought would point out that Jesus dined with publicans and other lost souls rather than with the clergy of the day. It is argued that development serves no end if concentration of extension efforts is on those who already have comparatively more opportunity and ability to make use of agricultural innovations. It does not serve to elevate the standard of living of those whose need is not as great as others. Those with a relatively favorable attitude toward innovations will by their own efforts progress somewhat, whereas those who acutely need extension will not progress at all. To carry extension to those in a position to best utilize it, the *latifundistas,* would be futile in many cases since it is often profitable for them to leave land underproductive. Aiding those who can best utilize extension would only serve to widen income disparities between rich and poor, adding to, not diminishing, social and political pressures in Latin America.

Dr. Dempsey Seastrunk, Assistant Director of Extension at Texas Agricultural and Mechanical University, stresses the need for extension to reach the "lowest of the low":

> It is better to take a lot of time to gain modest results with a large number of people than to use a little time to get large results with a few people. The key idea is to understand the value and the goals of the target group and work gradually but continually with each member until modest success is achieved. Once the end of the beginning is reached, progress can be more rapid. As short-range goals are reached, longer-range and more substantial goals can be set and worked toward. Getting better before getting bigger is the basic philosophy to use (SRC, 1977:3).

On the other hand, resources are limited, and the type of progress posited by Dr. Seastrunk moves with glacial speed. Dr. Fernando del Rio speaks for those who would maximize more immediate gains from extension:

> Extension must attend first to farmers who can make best use of teaching of extension and are in the best condition to augment production and income as an indispensable step to arriving at the ideal of working with everyone. Extension cannot pretend to assist a population without economic means, without sufficient lands, and with a very low education level; a population which would be better served by social programs (IICA, 1965:18).

He continues:

> I believe that the moment has come in which the extension services must define the population it is going to serve and fix attention on priorities for the needs of the selected population. In my opinion, in the Latin American context, it is the upper and middle class farmers, especially the latter, which presently constitute the farming groups which can make best use of the teaching of agricultural extension (IICA, 1965:32).

In the evaluation of this writer, cost-benefit analysis is applied inappropriately by Dr. del Rio. He ignores those benefits from extension that cannot be given a monetary value. The extension services are not running a business for profit. They are using public funds for the public good, and there can be little good in exacerbating the social cleavage in rural Latin America. Resources may be at a premium, but glacial speed is better than no speed and the latter velocity is exactly where Dr. del Rio would leave the *campesino*.

THE U.S. ROLE

Whatever choices are made in Latin America about the questions what, how, and to whom, it is certain that the choices will be influenced by the attitudes and the actions of their neighbors to the north. America's impact on Latin American extension has been and will be felt in four ways:

First, in an effort to consolidate hemispheric unity during and after World War II, the United States undertook to assist most Latin American countries in establishing extension services. Most countries had no agencies like this prior to World War II. There existed an occasional Bureau of Fomento, but nothing in the educational line. U.S. teams sent to organize the services were composed of a chief extension specialist, a rural youth specialist, and a home economist. This triumvirate is reflected today in that most services have, in addition to their agricultural information programs, rural youth organizations akin to 4-H clubs (4-S in Spanish) and home economics programs directed toward the *amas de casa*. After the initial organization and development stages, the missions were transferred to local control with Americans remaining as advisers. The imprint of the Americans on extension is indelible. Recruited from state extension services where many had been professors or county agents, their emphasis was on extension systems and methods, rather than agricultural content. This is not to say that no technical aid was given. Foresters, agronomists, and the like were from time to time posted with the extension service. After the U.S. gave up control of extension, the services suffered from lack of funds and support from indigenous governments. Poorly paid agents and ill-repaired jeeps were the result. The product of extension is likely to have suffered as well. The U.S. presence continues in the form of the Agency for International Development, the Inter-American Development Bank, the Inter-American Institute for Agricultural Sciences, and projects supervised by various American universities (Rice, 1974:53-100).

Second, there is the influence that America has as a market for Latin American agricultural products such as coffee and bananas. As such, it tends to make the task of extension more difficult by supporting export rather than consumption agriculture. This role would also tend to support a productionist view of extension.

Third, the United States acts as a training center for agricultural scientists and administrators from Latin America. Trained at the land-grant colleges, these foreign students take land-grant attitudes back to their own countries. Dr. Leonard Otto of the Agency for International Development elaborates: "In essence, the trained participant usually uses his scientific training for a few years or months. Sooner or later he is thrust into the crunch of government planning and administration and is frequently in and out of these activities as the political winds change" (SRC, 1977:2). Thus, as these students progress from scientific to administrative functions, attitudes learned in American agriculture colleges become ingrained in Latin American agricultural planning and administration.

Fourth, the United States serves as a center of research, development, and marketing of agricultural technological innovations. Public funds for agricultural research must be spent with domestic farmers in mind, so the research in tropical and labor-intensive agriculture is left to private entities, usually multinational corporations who do such research when it is profitable. Most American agricultural technology is developed for the high-capital, labor-scarce, large-volume, agribusiness-type American market. It is said to be no problem to design low-cost, labor-intensive technology to meet Latin American conditions; but the risks of selling to a market where most farmers are poor, without access to credit, and without large enough land holdings to use it efficiently are greater than the risks of selling existing technology to those who will buy it. The end result is that the technology pushed by American corporations is a profitable investment only for the rich, thereby serving to perpetuate the status quo in agriculture south of the border. Yudelman, Butler, and Banerji affirm this belief:

> The bulk of research and development with respect to the manufactured inputs used by farmers is undertaken in the private sector of developed countries. The firms that undertake this research are profit-oriented and their most important markets are in the developed countries where rural incomes are highest. . . . As a result little emphasis has been placed on meeting the specific needs of hundreds of millions of farmers who have low incomes and whose production requirements do not coincide with those of the commercial producers by virtue of the combination of their location, the nature of their output, and the scale of their operations (1971:146-58).

Thus, American research and corporate might serve not only to preserve existing agricultural structures in Latin America, but inhibit the development of innovations which would be useful to the *campesino*.

CONCLUSION

Latin America's extension services are its first-line defense against increasing poverty, overpopulation, and social inequities. How well extension is carried out not only will affect the rural areas but will increase or stay the flood of jobless, unskilled, impoverished rural people swelling the *barrios pobres* of the cities. How well the *extensionista* does his job can determine to a large extent whether Latin America reaches the twentieth century by the twenty-first or continues in cycles of tyranny, hunger, and intermittent bloodletting.

To do their critical task properly, the extension services must (1) opt for a people-oriented extension emphasis, (2) use basic education in conjunction with, not in opposition to, *fomento*, (3) assign the highest priority to working with the poorest and least capable, and (4) create indigenous agricultural research capacity and resist the siren song of inappropriate technology from the United States and other developed countries. These ideal solutions, however, fly in the face of the political and power realities of Latin America. If these are the solutions, then it must follow that current political and power structures are part of the problem. Any success in extension will depend upon whether these power realities can be changed.

It is for this reason that the four decisions facing Latin American extension are part of a larger decision that must be made: What is to be the role of extension in overall development policy? Extension is not a neutral process. It can be used as an instrument of status quo maintenance or as an instrument of development. Extension, with its potential for the spread of attitude change, could even be used as a means of resisting existing power structures. Moreover, whether extension is used to perpetuate the problems of under-development or is used as an instrument of change is, in part, a decision that must be made by the extension agent himself.

REFERENCES

Food and Agricultural Organization. 1971. *La Extensión Rural en America Latina y el Caribe.* Rome: FAO.
Inter-American Development Bank. 1967. *Agricultural Development in Latin America.* Washington, D.C.: IDB.
Instituto Interamericano de Ciencias Agricolas. 1965. *Primer Seminario Latinoamericano de Profesores de Extensión Agricola.* Turrialba, Costa Rica: IICA.
Lappé, Frances Moore, and Joseph Collins. 1977. *Food First: Beyond the Myth of Scarcity.* Boston: Houghton Mifflin Company.
Lipton, Michael. 1977. *Why Poor People Stay Poor.* Cambridge, Mass.: Harvard University Press.
Lopez Cordovez, Luis A., and Juan Galecio Gomez. 1961. *Informe y Manual de Extensión Agricola.* Quito, Ecuador: Junta Nacional de Planificatión y Coordinación Economica Misión, F.A.O. en el Ecuador.
Rice, Edward B. 1974. *Extension in the Andes.* Cambridge, Mass.: MIT Press.
SRC. 1977. *Southern Regional Conference on Training of International Students in Agriculture Summary Report.* Lexington, Ky.: University of Kentucky (mimeo).
Whyte, William F. 1977. Potatoes, Peasants, and Professors: A Development Strategy for Peru. *Sociological Practice* 2:7-23.
Yudelman, Montague; Gavan Butler; and Ranadev Banerji. 1971. *Technological Change in Agriculture and Employment in Developing Countries.* Paris: OECD.

Index

"Satellite" towns (*see* Coal industry, coal towns)
Schapera, Issac, 159
Schatz, Sayre P., 161
Schmitz, Andrew, 3
Schroyer, Trent, 83, 92, 104–5
Schuh, G. Edward, 136
Schultz, Theodore W., 146
Schumacher, Ernst F., 106, 151
Schutz, Alfred, 125
Science: assumption, 98; botanical gardens, 132–33; changing role, 116–17; commodification, 97, 117; comparison of medicine and agriculture 88–104; complementarity principle, 116; externalist position, 113–17; hegemonic system, 93–100, 105; internalist position, 113–17; legitimacy, 93–97, 117; problem of unity, 116; pure/applied science, 91–93; reductionism, 102–4; support of research, 93–95; world view, 85–88
Scientific journal articles: for selected cereals, *figure,* 150; for selected export crops, *figure,* 151
Scovell, Melville, 46
Seastrunk, Dempsey, 175
Seidel, Robert, 149
Shackelford, Nevyle, 34, 56, 58, 61
Shepard, Paul, 100
Sherman Anti-Trust Act, 71–72, 76–78
Shifting cultivation, 160
Shoemaker, F. Floyd, 61, 142
Shryock, Richard H., 85, 94
Sigurdson, Jon, 152
Sillery, Anthony, 158
Smith, Adam, 11, 116
Smith, J. Allan, 43–48, 56–57, 62, 97
Smith, J. Russell, 51–52
Smith-Lever Act, 47, 51, 60, 73, 96
Snodgrass, Milton M., 19–20
Social sciences, entry into agricultural research, 142–144
Soil Conservation and Domestic Allotment Act, 79
Soil Conservation Service, 174
Sorrenson, M. P. K., 141
Southern Agriculturalist, 49, 54
Spicer, Edward H., 162–63
Spitz, P., 136
Sprague, G. F., 152
Sprague, H. G., 98
Starr, Paul, 85, 90, 103
"State": definition, 69–70; intervention: class-interest perspective, 70–71/ pluralist perspective, 70/ tobacco products, Kentucky, 75–80/; role in capitalist nations, 69–71
Strauss, Anselm, 118
Strickland, Stephen P., 94

Subsistence farming, 132, 160
Suchman, Edward A., 164
Sumner, William Graham, 72
Sutton, S. B., 72
Systems analysis, 149–52

Taussig, Michael T., 97, 100
Technical Conference on Agricultural Extension and Rural Youth, 168
Technocratic strategy: agriculture, 85, 93–100, 105; definition, 83; medicine, 85, 93–100
Tennant, Richard B., 79
Texas Agricultural and Mechanical University, 175
Thomas, John D., 159
Timber industry, Kentucky, 28, 32–34
Tipps, Dean C., 162, 164
Tobacco and Health Research Institute, 121
Tobacco: cooperatives, 76, 78; research, 77; products, Kentucky: farmers' problems, 78–79/ KAES relationship, 74–80, 96/; legislation, 79; scientists, 121
Tobacco Institute, 75
Tobacco Trust, 45
Transportation, farmers' problems, 15
Transylvania University, 43–44
True, A. C., 133, 134
Tudiver, Sari, 61
Turbayne, Colin Murray, 84, 105
Turner, Wallace B., 31, 44, 48–49
Tuten, Robert T., 101
Tyler, Gus, 19, 21–22

"Undiluted goods" in progress of science, 2, 149
United Nations, 139
United Nations Development Program (UNDP), 146
United Nations Research Institute for Social Development (UNRISD), 152
U. S. Coal and Coke Company, 35
United States Department of Agriculture (USDA), 11, 14, 19–21, 30, 46–47, 51–52, 57–58, 78, 105, 123, 133–35
University of Chicago, 143
University of Kentucky College of Agriculture: establishment of, 43–48; Home Economics Department, establishment of, 46; John Bowman, 44; policies toward farming, 42–49; President Patterson, 45; W. A. Kellerman, 45
University of Kentucky Cooperative Extension Service (*see* Cooperative Extension Service, Kentucky)
Uphoff, Norman, 143

Vallianatos, E. G., 151
Varela, Louis Quiros, 167

About the Contributors

Lawrence Busch is an associate professor of sociology in the College of Agriculture at the University of Kentucky. He has extensive field experience in the United States and West Africa relating to the problems of creating effective agricultural research institutions. He has published articles in a variety of journals on the agricultural sciences, science policy, the sociology of science, and related questions of social change.

Christopher Dale is a Ph.D. candidate in sociology at the University of Kentucky. He is currently visiting assistant professor of sociology at SUNY Plattsburgh. His interests include land-tenure patterns and health care in the United States.

Thabo Fako is currently teaching sociology at the University College of Botswana in Gaborone. His interests include the sociology of development and health delivery systems in sub-Saharan Africa.

Donald K. Kazee has a master's degree from the Patterson School of Diplomacy and International Commerce of the University of Kentucky. He is currently a third-year student at Georgetown University Law Center.

William B. Lacy is an associate professor of sociology in the College of Arts and Sciences at the University of Kentucky. He has conducted social psychological research on the United States public agricultural research and extension system. He has published in journals in sociology, social psychology, and education on socialization, values, and social psychology of the agricultural sciences.

Sally Maggard is a Ph.D. candidate in sociology at the University of Kentucky. She has been active in the Council of the Southern Mountains and as editor of *Mountain Life and Work*. She is particularly interested in problems of social development in Appalachia.

S. Buik Mohammadi is a Ph.D. candidate in sociology at the University of Kentucky. His interests include social theory and social change. His dissertation is concerned with the recent revolution in Iran.

S. Randi Randolph is a Ph.D. candidate in sociology at the University of Kentucky. Her interests include the treatment of women by health-care professionals in the United States. Her dissertation focuses on improving the quality of interaction between pregnant women and their physicians.

Carolyn Sachs is a Ph.D. candidate in sociology at the University of Kentucky. She has published papers on the quality of public services and domestic agricultural science policy. Her dissertation concerns the changing role of American farm women over the past century.